Emerging Technologies in Telecommunications

Matthew N. O. Sadiku, Ph.D., P.E.

Regents Professor Emeritus and IEEE Life Fellow
Prairie View A&M University
Prairie View, TX 77446
Email: sadiku@ieee.org
Web: www.matthew-sadiku.com

BOOK FILMS MEDIA

Date published: *February 14, 2025*

Matthew N. O. Sadiku

sadiku@ieee.org
www.matthew-sadiku.com

Paperback: *979-8-9912266-6-0*
eBook: *979-8-9912266-7-7*
Hardcover: *979-8-9915027-0-2*

BookFilmsMedia

2780 South Jones Blvd Suite 200 - 4007
Las Vegas, NV 89146 United States
+1 725 - 238 - 6534

Cover credits: Photo by Rasheeque Ahnaf (Piash) on Unsplash.
@the_ahnafpiash

DEDICATED TO MY WIFE:

JANET O. SADIKU

BRIEF TABLE OF CONTENTS

PREFACE

The telecommunication is everywhere, and we rely on it evermore throughout our daily lives. It connects people around the globe. Telecommunications serve as the backbone of modern connectivity. The telecommunications sector keeps the world connected. Everything from work to play is shaped by modern telecom's unified voice, video, and information architecture. The modern telecommunications ecosystem is a complex beast. Its tentacles reach far and wide, affecting everything from mundane daily tasks to global business operations. Our economies and entire business infrastructures are built on modern telecoms.

Emerging technologies are shaping our societies. They continue to affect the way we live, work, and interact with one another. They take center stage as a critical force for digital transformation. The telecom industry faces challenges like managing network demands, optimizing infrastructure, resolving customer issues, and strengthening security. Emerging technologies are helping telecom companies address these issues and improve efficiency. An emerging technology is the one that holds the promise of creating a new economic engine and is trans-industrial.

Telecommunication technologies have become an inevitable part of our day-to-day lives, from mobile connectivity to home connectivity, office conferencing to remote surgeries. In today's digital age, the telecom industry has become a crucial pillar for global connectivity and communication. The industry is at the forefront of the digital transformation. Telecom giants are leveraging the goodness of modern technologies to boost overall operations and multiply customer experience. These new technologies digitally transform the telecom sector to meet the ever-growing demands of businesses and customers and to stay competitive in the market.

This book examines emerging technologies and their impact on the telecommunications industry. It is organized into twelve chapters that summarize emerging technologies and their applications in telecommunications.

Chapter 1 - Introduction

This chapter considers some of the leading emerging technologies within the telecommunications sector and serves as an introduction to the book. It discusses some emerging technologies in telecommunications such as artificial intelligence, cloud computing, edge computing, drones, big data, Internet of things, blockchain, 5G networks, 3D printing, virtual reality, augmented reality, and cybersecurity. The convergence of all these emerging technologies in telecommunications industry is setting the stage for a transformative era in telecommunications. Together, all these technologies will help in making communication faster, smarter, and greener.

Chapter 2 - Artificial Intelligence

This chapter examines the roles of artificial intelligence (AI) and generative AI in the telecommunications industry. Artificial intelligence refers to the ability of a computer system to perform human tasks (such as thinking and learning) that usually can only be accomplished using human intelligence. AI is transforming industries across the globe and telecommunications industry is no exception. Integrating AI in the telecoms industry is a continuous and evolving process with the potential to enhance efficiency, improve customer experiences, and drive innovation across various aspects of telecommunications. AI technology is here to stay and will continue to significantly impact the telecommunications industry.

Chapter 3 - Drones

This chapter examines the applications of drones in the telecommunications industry. Drones, also known as unmanned aerial vehicles (UAVs), may be regarded as any aircraft system without a flight crew onboard. Drones have evolved in the past few years to become exceptionally versatile across practically every industry. As drones in telecommunications industry become increasingly prevalent, they are transforming the way companies approach tower inspections, network expansions, and emergency response. The future of telecom drones looks bright and promising as drone technology continues to evolve and improve.

Chapter 4 - Big Data

This chapter examines the roles of big data and big data analytics in telecommunications. Big data applies to data sets of extreme size (e.g. exabytes, zettabytes) which are beyond the capability of the commonly used software tools. We are in the big data age, where businesses in every industry must deal with vast volumes of data. Big data and the technologies it relies on, such as AI and machine learning, are the driving force behind progress in the telecommunications industry. Big data in the telecom industry encompasses a wide range of information, including customer profiles, call records, network logs, location data, social media interactions, and more. It is revolutionizing the telecommunication industry by enabling targeted marketing and enhancing customer behavior insights.

Chapter 5 - Internet of Things

This chapter examines the roles of Internet of things (IoT) in the telecommunications industry. The Internet of things is a concept that refers to a network of everyday physical objects or devices that are connected to the Internet and can communicate with each other and with end-users. Telecommunications play a key role in creating an interconnected world through the Internet of things. They provide the essential infrastructure and services needed to connect IoT devices to each other, to the Internet, and to other systems. Integration of the IoT and telecommunication industry will become a solid competitive edge that optimizes productivity, improves service quality, unlocks extra opportunities, raises customer loyalty, increases profits, and leads the telecom business to success.

Chapter 6 - Cloud Computing

This chapter discusses the roles of cloud computing in the

telecommunications industry. Cloud computing is a means of pooling and sharing hardware and software resources on a massive scale. It is the provision of scalable computing resources as a service over the Internet. It has become a game-changer, empowering telcos to embrace agility, efficiency, and innovation like never before. Cloud computing in telecom is a software defined infrastructure that allows telcos to store and process data remotely in the data processing centers.

Chapter 7 - Edge Computing

In this chapter, we explore the applications of edge computing in the telecommunications industry. Edge computing calls for processing the data at the edge of the network, as opposed to a data center or cloud. This approach offers reduced latency, improved scalability, and enhanced security. In the ever-evolving landscape of telecommunications, the emergence of edge computing stands as a beacon of promise, offering ultra-low latency solutions that propel networks into the future. Its most significant advantage in telecom is the drastic reduction in latency. Edge computing is poised to shape the future of telecommunications and usher in a new era of connectivity and innovation.

Chapter 8 - Blockchain

In this chapter, we explore the potential of blockchain in the telecommunications industry. Blockchain, a type of distributed digital ledger technology, is a relatively new and exciting way of recording transactions in the digital age. It is a chain of blocks in which each block contains a list of transactions. Blockchain is changing how telecom works, offering better security, efficiency, and trust. Although blockchain technology is relatively new to the telecommunications industry, it has the potential of revolutionizing telecommunications in the future.

Chapter 9 - 5G Network

This chapter examines the applications of 5G networks in telecommunications. 5G stands for the fifth generation of cellular networking that promises to deliver faster speeds, lower latency, and greater reliability than its predecessors. The introduction of 5G technology is set to revolutionize the way we communicate, work, and live. The capabilities of 5G have opened opportunities for the telco industry to serve customers closer to the edge of the network. The future of 5G in the telecommunications industry is highly promising and holds immense potential to revolutionize multiple industries including healthcare, transportation, manufacturing, entertainment, and telecommunications.

Chapter 10 -3D Printing

The chapter covers the applications of 3D printing in the telecommunication industry. 3D printing is a digital manufacturing technology that can create physical objects from a digital file like CAD. This technology is growing and can be used for small productions of these computer-made designs in many industries, including in the telecom industry. With the ability to create complex parts and products quickly and efficiently, 3D printing is transforming the way that telecom companies operate. Although the use of the 3D printing technology is not prevalent in the telecommunications sector, this field is considered as a potential area of growth of the industry.

Chapter 11 - Virtual and Augmented Reality

This chapter explores the immersive technologies and their vital roles in telecommunications. Virtual reality (VR) is a computer technology that creates a simulated environment, while augmented reality (AR) overlays digital information onto real-world objects. These are immersive technologies that can redefine how we interact with the digital world. They are rapidly transforming various industries, including telecommunications. The telecommunications industry has seen a gradual adoption of AR/VR technology, with companies exploring its potential for enhancing communication and collaboration among teams.

Chapter 12 - Cybersecurity

In this chapter, we discuss how cybersecurity measures are applied in the telecommunications industry. Cybersecurity refers to the protection of a computer system and/or network from attack as well as the risk of losing data or information, unauthorized access, and fraud. Telecom operators store various data, from social security numbers to credit card details. This wealth of sensitive data makes large-scale telecom firms appealing targets for bad actors. By prioritizing cybersecurity, telecom companies will not only safeguard their networks but also reinforce trust, enabling a future where seamless, secure communication remains a cornerstone of our interconnected world.

This is a comprehensive introductory text on the issues, ideas, theories, and applications of emerging technologies in the telecommunications industry. It provides an overview of each technology in simple terms so that beginners can understand it. It is a must-read book for anyone who wants to learn about modern telecommunications, which have become vital in many areas of life.

I am grateful for the support of Dr. Annamalia Annamalai, the department head of the Department of Electrical and Computer Engineering, and Dr. Pamela Obiomon, the dean of the College of Engineering at Prairie View A&M University, Prairie View, Texas. Special thanks are due to my wife Dr. Janet Sadiku for helping in various ways.

- M. N. O. Sadiku

ABOUT THE AUTHOR

Matthew N. O. Sadiku received his B. Sc. degree in 1978 from Ahmadu Bello University, Zaria, Nigeria and his M.Sc. and Ph.D. degrees from Tennessee Technological University, Cookeville, TN in 1982 and 1984 respectively. From 1984 to 1988, he was an assistant professor at Florida Atlantic University, Boca Raton, FL, where he did graduate work in computer science. In total, he received seven college degrees. From 1988 to 2000, he was at Temple University, Philadelphia, PA, where he became a full professor. From 2000 to 2002, he was with Lucent/Avaya, Holmdel, NJ as a system engineer and with Boeing Satellite Systems, Los Angeles, CA as a senior scientist. He is presently a Regents professor emeritus of electrical and computer engineering at Prairie View A&M University, Prairie View, TX.

He is the author of over 1,330 professional papers and over 140 books including "Elements of Electromagnetics" (Oxford University Press, 7th ed., 2018), "Fundamentals of Electric Circuits" (McGraw-Hill, 7th ed., 2020, with C. Alexander), "Computational Electromagnetics with MATLAB" (CRC Press, 4th ed., 2019), "Principles of Modern Communication Systems" (Cambridge University Press, 2017, with S. O. Agbo), and "Emerging Internet-based Technologies" (CRC Press, 2019). In addition to the engineering books, he has written Christian books including "Secrets of Successful Marriages," "How to Discover God's Will for Your Life," and commentaries on all the books of the New Testament Bible. Some of his books have been translated into ten languages: French, Korean, Chinese (and Chinese Long Form in Taiwan), Italian, Portuguese, Spanish, German, Dutch, Polish, and Russian.

He was the recipient of the 2000 McGraw-Hill/Jacob Millman Award for outstanding contributions in the field of electrical engineering. He was also the recipient of Regents Professor award for 2012-2013 by the Texas A&M University System. He is a registered professional engineer and a life fellow of the Institute of Electrical and Electronics Engineers (IEEE) "for contributions to computational electromagnetics and engineering education." He was the IEEE Region 2 Student Activities Committee Chairman. He was an associate editor for IEEE Transactions on Education. He is also a member of Association for Computing Machinery (ACM). His current research interests are in the areas of computational electromagnetic, computer science/networks, engineering education, and marriage counseling. His works can be found in his autobiography, "My Life and Work" (Author's Tranquility Press, 2024) or his website: www.matthew-sadiku.com. He currently resides with his wife Janet in Westlake, Florida. He can be reached via email at sadiku@ieee.org

TABLE OF CONTENTS

CHAPTER 1
INTRODUCTION

"It has become appallingly obvious that our technology has exceeded our humanity."

—Albert Einstein

1.1 INTRODUCTION

Emerging technologies take center stage as a critical force for digital transformation. The technology domain is continuously modernized with innovations and has immense potential to create intelligent solutions for the future. The emergence of artificial intelligence (AI), 5G/6G, the Internet of things (IoT), big data, and more have marked incredible transformations in various industries. And the telecom industry is no exception.

Telco leaders are broadly aware of the magnitude of transformation that the moment demands. The telecom industry faces challenges like managing network demands, optimizing infrastructure, resolving customer issues, and strengthening security. Emerging technologies are helping telecom companies address these issues and improve efficiency. Figure 1.1 shows how telecom is being revamped by emerging technologies [1].

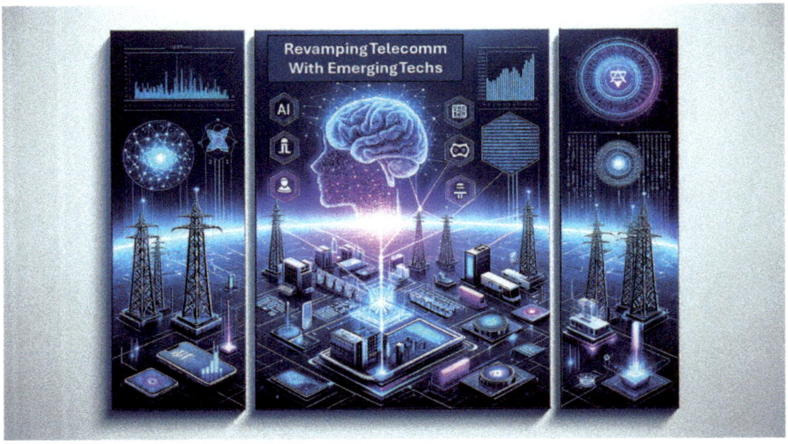

Figure 1.1 Telecom is being revamped by emerging technologies [1].

The pace and expanse of digitalization have increased exponentially in the past few years. The concept of digital transformation is driving the changes we see in the world around us. Telecommunication industry is the pioneering sector of industrial technological transformation, noted for its complex infrastructure, as depicted in Figure 1.2 [2].

Figure 1.2 Telecommunication industry is noted for its complex infrastructure [2].

Digital transformation offers telecom industry an opportunity to reconstruct their market position, re-create their business systems, and produce inventive offerings for customers. From the first telephone to today's high-speed Internet and mobile networks, telecom has consistently pushed the boundaries of what's possible in communication. The telecom sector forms the backbone of the technologies required to increase the scope of digitalization. Today, as the world enters the era of the fourth Industrial Revolution, the responsibility of a smooth transition to a "smart", connected world falls on the telecom sector. The emergence of new, big-name players to challenge industry stalwarts like AT&T, Verizon, and Comcast is driving growth [3].

Telecommunications serve as the backbone of modern connectivity. The industry is at the forefront of the digital transformation. Telecom giants are leveraging the goodness of modern technologies to boost overall operations and multiply customer experience. These new technologies digitally transform the telecom sector to meet the ever-growing demands of businesses and customers and to stay competitive in the market. The emergence of the new technologies in telecommunication is a revolution unfolding before our eyes. The telecommunications industry stands on the brink of a significant transformation driven by the emerging technologies [4].

This chapter considers some of the leading emerging technologies within the telecommunications sector and serves as an introduction to the book. It begins with describing what emerging technologies are. It discusses some emerging technologies in telecommunications. It highlights the benefits and challenges of these technologies. It concludes with some comments.

1.2 WHAT ARE EMERGING TECHNOLOGIES?

Technology may be regarded as a collection of systems designed to perform some function. It can help alleviate some of the challenges facing business today. Emerging technology is a term generally used to describe new technology. The term often refers to technologies currently developing or expected to be available within the next five to ten years. Any imminent, but not fully realized, technological innovations will have some impact on the status quo.

Emerging technologies are shaping our societies. They continue to affect the way we live, work, and interact with one another. Emerging technology (ET) lacks a consensus on what classifies them as "emergent." It is a relative term because one may see a technology as emerging and others may not see it the same way. It is a term that is often used to describe a new technology. A technology is still emerging if it is not yet a "must-have" [5]. An emerging technology is the one that holds the promise of creating a new economic engine and is trans-industrial. ET is used in different areas such as media, healthcare, business, science, education, or defense.

The characteristics of emerging technologies include the following [6]:

- *Novelty:* Emerging technologies are typically new or novel, meaning they have yet to be widely adopted or used. They often represent a significant departure from existing technologies or processes.
- *Potential for Disruption:* Emerging technologies have the potential to disrupt existing markets, industries, or ways of doing things. They may also displace existing businesses or industries.
- *Uncertainty:* Because emerging technologies are still in the early stages of development, there is often a high uncertainty surrounding their future potential and impact. It can be challenging to predict how they will evolve.
- *Rapid Change:* Emerging technologies often evolve rapidly, with new developments and innovations emerging frequently. It can make keeping up with the latest trends and advancements challenging.
- *Interdisciplinary:* Emerging technologies often involve multiple disciplines or fields of study, such as computer science, engineering, and biology. They may require collaboration across different fields and industries to develop their potential fully.

Emerging technologies are worth investigating. They are

responsible for developing new products or devices. The military often looks to emerging technologies for new services or tools that will help them create a competitive business advantage. As emerging technologies continue to evolve, engineering is poised for a transformative future. Emerging technologies have driven innovation and progress in today's rapidly evolving digital landscape. The collective impact of emerging technologies such as artificial intelligence, machine learning, cloud computing, big data, and the Internet of things is undeniably transformative.

1.3 EMERGING TECHNOLOGIES IN TELECOMMUNICATIONS

The telecom industry is growing rapidly and becoming one of the leading sectors of the global economy. The emerging technologies are likely to impact business and marketing activity in the coming years. With the advent of these technologies, the telecom industry is witnessing a shift towards efficient telecommunications network management and infrastructure sharing. Emerging technologies in the telecommunications industry include the following [7,8]:

1. *Artificial Intelligence:* Artificial intelligence (AI) and automation are the most transformative and consequential technologies to impact telecommunications since 5G. These two powerful technologies have revolutionized almost every sector and telecom is no exception. An area where AI and automation are helping telecom companies is customer service. Besides this, AI and automation are also critical for network security and operations management. Automation solutions based on artificial intelligence (AI) and machine learning (ML) are becoming popular as telecom companies try to improve efficiency and reduce costs. AI and ML stand out as significant trends in the telecom industry, exerting a profound impact. AI is currently working to transform the telecommunications sector. Advances in AI are unlocking opportunities for organizations at every point along the value chain. Telcos can use AI to optimize networks, proactively address maintenance issues, and minimize churn. The release of ChatGPT, Open AI chatbots has cut down the human task to generate high-quality content based on the trained data. Generative AI is

winning the public's desire. Generative AI can transform customer experience by supplying customers with highly personalized content, offers, and proactive service-related outreach based on usage patterns, purchase history, and other considerations. By using generative AI to simulate sophisticated cyberattacks, operators can identify vulnerabilities and enhance network resilience. Gen AI in next-generation networks facilitates proactive, predictive, and preventive maintenance. Nokia harnesses AI to power ahead digital transformation efforts. See Figure 1.3 [9].

Figure 1.3 Nokia harnesses AI for digital transformation efforts [9].

2. *Cloud Computing:* Cloud computing is a rapidly growing trend in the telecom industry, primarily driven by the widespread use of IoT devices and the deployment of advanced ML algorithms. In this landscape, cloud startups play a crucial role by providing enhanced connectivity and integration solutions across equipment, platforms, infrastructure, and various company functions. Scalability is one of the key benefits of cloud computing, enabling telecom operators to scale up and down based on the fluctuating demands of the consumers. Cost is an added advantage for telecom operators when they migrate to the cloud as the expenditures shift from CapEx to OpEx. As 5G network requires more flexible networks, telecoms need to engage in cloudification to optimize for lower costs, more elasticity, and increased speed. As illustrated

in Figure 1.4, 70 percent of Telecom workload is predicted to be hosted in the cloud by 2025 [8].

Figure 1.4 70% of Telecom workload is predicted to be hosted in the cloud by 2025 [10].

3. *Edge Computing:* Although edge computing is considered an extension of cloud computing, edge computing requires a fully distinct deployment and development ecosystem. As data generation skyrockets, edge computing has become crucial in telecom for processing data closer to the source. Edge computing in the telecom sector is a distributed computing framework that brings enterprise applications closer to data sources. In other words, it involves moving data computation and storage closer to the data source, as opposed to relying on a centralized remote cloud. This reduces latency and bandwidth use, enabling real-time data analysis and faster decision-making. Edge computing is a fundamental shift in how we utilize data. This approach reduces latency, enhances bandwidth, streamlines maintenance, and enables companies to expand their computing capacity more affordably by integrating devices with edge data centers. Edge computing allows for real-time data processing, which will unlock use cases across industries—from remote healthcare treatment to remote management of mining operations to sustainability solutions. As telcos embrace edge computing, they will face

rising costs from energy consumption, network maintenance, and investments associated with reconfiguring network backhaul and backbone. Edge computing significantly enhances the end-user experience quality and reduces data exchange demands. AT&T aims to use edge computing to improve performance. See Figure 1.5 [9]. Using edge computing, leading provider AT&T aims to bring data processing closer to the source to reduce latency and improve performance.

Figure 1.5 AT&T aims to use edge computing to improve performance [9].

4. *Cybersecurity:* The communications industry is one of 16 critical infrastructure sectors whose assets, systems, and networks, whether physical or virtual, are considered so vital to the United States. Telecoms are thus a top target for cybercriminals. Cybersecurity and resilience are therefore significant concerns for the telecom industry. Cyberattacks and threats are increasing and becoming complex with technological advancements. The rising frequency of cyberattacks, coupled with the proliferation of low-security IoT devices and the emergence of new CPU hardware vulnerabilities, presents a formidable challenge in addressing security risks. Telecom operators are prioritizing cybersecurity measures to protect their networks, infrastructure, and customer data. To meet rising consumer expectations around digital trust, telcos should consider investing in cybersecurity solutions. Those that do so will position themselves to introduce new offerings by building digital identity services on next-generation networks and technologies. The future of cybersecurity in the telecom domain

is characterized by continuous adaptation and embracing cutting-edge technologies.

5. *5G Technology:* 5G is the 5th generation mobile network, as depicted in Figure 1.6 [9]. The 5G Networks, which promises to deliver faster speeds, lower latency, and increased capacity compared to its predecessors, are the latest innovation that telecom has adopted. Connectivity technologies are in a continuous state of evolution, encompassing both wired and wireless communication methods.

Figure 1.6 The 5th generation mobile network [11].

The expansion and increasing complexity of the Internet have raised the demand for high-speed connections with minimal latency. The adoption of 5G technology is one of the key trends in the telecom industry. 5G technology represents the upcoming major advancement in telecommunication networks and devices. Fifth-generation (5G) telecommunications infrastructure is dramatically expanding and improving connectivity, and sixth-generation (6G) infrastructure is poised to amplify this trend. Within the telecom industry, 5G introduces the capability for massive machine-type communications, facilitating the establishment of dense IoT networks. The market for 5G technology is expected to see substantial growth. Near-limitless connectivity will pave the way for new services like remote patient monitoring and next-generation customer experiences like virtual dressing rooms and conferences. Demand for connectivity is expected to increase further as customers seek out these innovative solutions. Connectivity aims to achieve global broadband coverage, as shown in Figure 1.7 [9].

Figure 1.7 Connectivity to achieve global broadband coverage [9].

Satellite constellations could help address lag issues in satellite communications and revolutionize telecommunications by providing Internet access in areas where fiber and tower infrastructure cannot be deployed. This has great potential to improve connectivity in rural areas. While 5G networks are still being rolled out, Huawei is already investing heavily in 6G research and development. 6G is predicted to support one million devices per square kilometer, compared to the 1,000 devices supported by 5G. One of the biggest developments in the post-pandemic period was the deployment of 5G services by various governments in their nations.

6. *Internet of Things:* IoT is a collection of devices connected to each other and the network. IoT comes with the concept that every device should be electronically integrated and interconnected. IoT has transformed the way we live, work, and interact with the world around us. IoT devices and sensors are significantly impacting various sectors of the technology economy, enhancing people's quality of life, boosting businesses' profitability, and improving overall management. IoT offers cost-saving benefits for governments seeking to reduce their IT-related expenses. The interconnected nature of devices, sensors, infrastructure, and computing components is opening up innovative avenues

for management. The industrial Internet of things (IIoT) enables more connected devices supporting new business models. The IIoT market is expanding as different industry verticals integrate IoT technology for smart manufacturing, healthcare, logistics, and more. The telecommunications industry has witnessed a growth in the adoption of connected devices, sensors, and semiconductors that form the framework for the Internet of things. The IoT is growing rapidly, and will eventually impact all industries and everyone's daily life.

7. *Virtual Reality:* Virtual reality (VR) and augmented reality (AR) technologies will continue to evolve in new ways, allowing telecom organizations to use all potential to profit from the opportunities they offer to create new customer experiences. While most AR/VR applications today support single-user, single-device static experiences, the startup's flexible cloud infrastructure makes content management and delivery simple. By making possible AR/VR products, deployable everywhere, the solution benefits brands and enterprises seeking to improve their consumers' experiences.

8. *Blockchain:* Blockchain's impact on telecommunications is profound, offering innovative solutions to longstanding security, transparency, and efficiency challenges. The technology solves the telecom industry's problems by providing decentralized, transparent, and tamper-proof accounting, allowing telecom operators to automate their billing processes. Using blockchain, telecom companies create smart contracts that automatically execute billing agreements based on pre-set terms, reducing the likelihood of human error and disputes. Blockchain could be the basis for developing innovative mechanisms for energy trading and gaining economic benefits by incorporating crypto-assets with future smart energy systems. Blockchain technology is redefining the foundations of telecommunications by bolstering security and ensuring unmatched transparency.

9. *Big Data:* Big data provides telecom companies with more insights into operations and customers and helps to promote growth while also increasing efficiency and profitability.

10. *3D Printing:* This is also known as additive manufacturing. It is transforming the design and manufacturing processes. This technology enables engineers to create complex, customized components with greater precision and efficiency. From rapid prototyping to on-demand manufacturing, 3D printing is disrupting traditional supply chains and reducing production costs. Industries such as aerospace, healthcare, and automotive are benefiting from this technology, as it allows for the creation of lighter, stronger, and more intricate structures.

1.4 BENEFITS

Digital transformation projects are making businesses work hard to remain competitive and best support customers. In the telecom sector, digital transformation is highly embraced to adopt innovation and revamp business strategies to stay competitive in the dynamic marketplace. Demonstrating that your telecom business is up to date with data privacy policies, security measures, and compliance with regulations helps build brand trust and confidence with customers. Other benefits include [1,12]:

- *Connectivity:* The demand for connectivity and networks has increased significantly as more and more consumers and businesses are seeking faster and high-performance connectivity. The telecommunications industry is expected to provide consumers with digital connectivity.

- *Automation:* Networks are becoming more and more advanced and complicated. Telecom companies are turning to advanced analytics and automation to manage and optimize their operations. Automation, powered by AI and ML, is streamlining network management tasks and other tasks. Telecom operators are adopting automation technology, multi-channel approach, and personalization services to create a great customer experience, which improves asset utilization. The future may see the rise of fully autonomous network operations managed by AI.

- *Sustainability:* Recognizing the environmental impact, the telecom industry in 2024 places a stronger emphasis on sustainability. Being one of the biggest contributors to global energy consumption and carbon emissions, the telecom industry is rapidly turning towards sustainable initiatives. The replacement of copper cables with fiber optic cables is also gaining traction as it significantly reduces energy consumption. Sustainable initiatives in the telecom industry will yield positive impacts such as reduced carbon footprint, lower operating costs, and improved brand image. Through sustainability initiatives, telecom companies can use less energy, reduce their carbon footprint, and help the planet. Looking ahead, sustainability is expected to remain a top priority for the telecom industry.

- *Renewable Energy:* Many telecom companies are deploying energy-efficient cooling systems, optimizing server utilization, and using renewable energy sources. Telecom companies are making significant investments in solar and wind farms to use renewable sources of energy for their operations. Emerging technologies are increasingly incorporated in the energy sector to facilitate both the design and operations of future smart energy systems with high penetrations of renewable energy and decentralized structures.

- *Cybersecurity:* The changing digital landscape has necessitated cybersecurity to become of critical concern. It plays a major role in industrial application, especially the Telecom providers who are not only responsible for protecting their own infrastructure but also for safeguarding the vast amounts of data they transmit.

- *Fraud Detection:* AI systems monitor network activities in real time to detect and prevent fraudulent actions, significantly enhancing security measures. T-Mobile, for instance, employs AI to analyze call patterns and identify potential fraud.

- *Personalized Services:* AI enables telecoms to offer customized services and content, enhancing customer satisfaction and loyalty, a strategy increasingly adopted by telecom providers to suggest tailored data plans and services.

- *Regulatory Compliance:* Data Science tools enable telecom companies to comply with various regulatory requirements by monitoring and analyzing data to ensure all operations are within legal bounds. This observation is crucial for maintaining operational legality and avoiding hefty fines.

- *Decentralization:* The Internet of things has also brought in a new level of communication based on predictive maintenance and decentralization. Many telecommunication companies have even developed solutions that capitalize on this decentralization aspect of this new-age technology.

- *Autonomous Vehicles:* The telecommunications industry has played a vital role in establishing infrastructure that can support the adoption of autonomous vehicles. The upgradation in telecommunication technologies has also been important for the birth and enhancement of telematics and navigation, which are crucial for the smooth functioning of autonomous vehicles.

1.5 CHALLENGES

As telecom companies embrace the new technologies, they must also address the associated challenges. Adapting to new technologies and market demands is critical to remaining competitive. By acquiring cybersecurity, data analytics, cloud computing, machine learning, and artificial intelligence skills, professionals can carve out a successful career in the emerging telecom industry. Other challenges include the following [13]:

- *Tech Talent:* To get ahead of these tech trends, telcos will need to develop long-term strategies for cultivating, attracting, and retaining the right talent, with the right skills. Long gone are the days when telcos were the employer of choice for technical talent. Telco operators with ambitious goals regarding diversity, equity, and inclusion should be particularly intentional about developing sustainable, long-term talent pipelines. As telcos evolve to deliver on the opportunities that emerging technologies unlock, they will need to be highly strategic about identifying and attracting talent with the expertise and abilities that each technology demands.

- *Data Quality:* With a growing range of data sources, the ability to profile users and target them with personalized content is expanding. However, ensuring the accuracy and reliability of data inputs while creating coherent user profiles is becoming increasingly critical.

- *Private Data Anonymization:* As telcos migrate to the cloud, protecting private and sensitive user data becomes paramount. Compliance with regional regulations like the GDPR necessitates anonymization. However, challenges arise during the migration process, especially with technology stack transitions, necessitating custom development and rigorous testing.

- *Data Governance:* With an influx of users and an expanding array of software accessing and modifying vast amounts of data, coherent data governance solutions are crucial. These solutions define and uphold standards and policies for data collection, storage, and processing. Meanwhile, telcos can leverage data insights to enhance their business operations and stay ahead of emerging trends in the telecommunications industry.

- *Regulations:* Regulations shape the deployment of new technologies. Regulations, privacy rules, and business models have yet to catch up to handle the accompanying data, whether transactional, electronic health records or behavioral tracking. Effective telecommunications regulation directly impacts economic growth. Policymakers should focus on regulations that facilitate the deployment of advanced technologies.

- *Space Debris:* This poses huge risks and safety threats to the existing space systems and activities since it may cause a collision with other space stations, satellites, or any other equipment. Governments, space agencies, and other institutions are funding or implementing their own initiatives to advance space sustainability, outside of multilateral efforts.

1.6 CONCLUSION

The telecom industry continues to be a beacon of digital transformation, integrating innovative technologies to enhance connectivity and communication. The convergence of all the above-mentioned emerging technologies in telecommunications industry is setting the stage for a transformative era in telecommunications. This technological synergy is poised to overcome traditional industry challenges, offering more reliable, secure, and user-centric services. Together, all these technologies will help in making communication faster, smarter, and greener.

In the coming years, with the help of these emerging technologies, telecom industries can renovate their operational and business models to enlarge their customer base and launch innovative products, solutions, and services. The future of telecom is bright and filled with exciting possibilities. More information about emerging technologies in the telecommunications industry can be found in the books [14-18] and the following related journal/magazines:

- *Transactions on Emerging Telecommunications Technologies*
- *International Journal of Telecommunications & Emerging Technologies*
- *Journal of Biomedical Systems & Emerging Technologies*
- *Mobile Magazine*

REFERENCE

[1] A. S. AlBalooshi, "Article 4: Revamping telecommunications with emerging technologies: AI, blockchain, & data science," April 2024,

https://www.linkedin.com/pulse/article-4-revamping-telecommunications-emerging-ai-data-albalooshi-ne0xf/

[2] "Comparative analysis of telecommunication regulation: USA vs. Africa and the role of emerging technologies,"

https://www.africanresearchers.org/comparative-analysis-of-telecommunication-regulation-usa-vs-africa-and-the-role-of-emerging-technologies/

[3] M. N. O. Sadiku, P. A. Adekunte, and J. O. Sadiku, "Telecommunications Industry: An Overview," *International Journal of Trend in Scientific Research and Development*, vol. 8, no. 6, November-December 2024, pp. 503-510.

[4] M. N. O. Sadiku, O. D. Olaleye, and J. O. Sadiku, "Emerging technologies in telecommunications," *International Journal of Trend in Research and Development*, vol. 11, no. 6, November-December 2024, pp. 134-139.

[5] M. Halaweh, "Emerging technology: What is it?" *Journal of Technology Management & Innovation*, vol. 8, no. 3, 2013, pp. 108-115.

[6] N. Duggal, "Top 18 new technology trends for 2023," July 2023,

https://www.simplilearn.com/top-technology-trends-and-jobs-article

[7] T. Lajous et al., "Tech talent in transition: Seven technology trends reshaping telcos," October 2023,

https://www.mckinsey.com/industries/technology-media-and-telecommunications/our-insights/tech-talent-in-transition-seven-technology-trends-reshaping-telcos

[8] "Discover the top 10 telecom industry trends in 2024,"

https://www.startus-insights.com/innovators-guide/top-10-telecom-industry-trends-innovations-in-2021/

[9] A. Jackson, "Top 10: Emerging technologies in telecommunications," August 2024,

https://mobile-magazine.com/top10/top-10-emerging-technologies-in-telecommunications

[10] K. Sharma, "9 Key trends in telecommunication to know in 2024," March 2024,

https://www.revesoft.com/blog/telecom/trends-in-telecommunication/

[11] "Everything you need to know about 5G,"

https://www.qualcomm.com/5g/what-is-5g

[12] "The future of telecom: Exploring emerging technologies and trends,"

https://www.allstarsit.com/blog/the-future-of-telecom-exploring-emerging-technologies-and-trends

[13] R. Makarchuk, "Top telecommunications industry trends shaping 2024 and beyond," July 2024,

https://intellias.com/top-telecom-trends/

[14] N. A. C. Pinoth, *Innovation and Emerging Technologies: Of Telecommunications*. Our Knowledge Publishing, 2024.

[16] A. Paktas and R. Komiya (eds.), *Virtual Reality Technologies for Future Telecommunications Systems*. Wiley, 2022.

[17] S. McMaster, The Telecommunications Industry (Emerging Industries in the United States). Greenwood, 2002.

[18] U. Vanka, *The Future Is BIG: How Emerging Technologies are Transforming Industry and Societies*. Business Expert Press,

2023.

[19] US Congress, *Emerging Telecommunications Technologies*. U.S. Government Printing Office, 1990.

CHAPTER 2
ARTIFICIAL INTELLIGENCE IN TELECOMMUNCATIONS

"We must address, individually and collectively, moral and ethical issues raised by cutting-edge research in artificial intelligence and biotechnology, which will enable significant life extension, designer babies, and memory extraction."

—Klaus Schwab

2.1 INTRODUCTION

Previously, telcos primarily offered basic telephone, Internet, and mobile network services. No longer limited to providing basic phone and Internet service, the telecom industry is at the epicenter of pushing the wide adoption of cutting edge technologies, led by mobile and 5G broadband services in the Internet of things (IoT) era. The global mobile data traffic is expected to multiply fivefold before the end of the decade. The telecom companies are struggling to manage this surge, grappling with network congestion, maintenance challenges, and heightened customer expectations for seamless connectivity.

In an increasingly competitive market, telcos must exert great effort to meet customer expectations, as numerous alternative service providers are available for consumers. To ensure that Telcos are meeting customers' service delivery and experience needs, Telcos have turned to AI technologies. In recent years, the telecommunication industry has experienced significant changes and a shift in focus towards innovation, with artificial intelligence (AI) playing a central role.

Telecommunications is an industry that serves over 5 billion customers worldwide, noted for its complex infrastructure, as depicted in Figure 2.1 [1]. Telecom providers must deal with complex network infrastructures that require constant optimization and maintenance. The growing complexity of network infrastructure and tens of billions of connected devices make it impractical and beyond human capacity to manage and operate networks using only traditional network management methods. AI and automation are helping networks overcome this immense challenge. The importance of AI in telecom is displayed in Figure 2.2 [2].

Figure 2.1 Telecommunications is noted for its complex infrastructure [1].

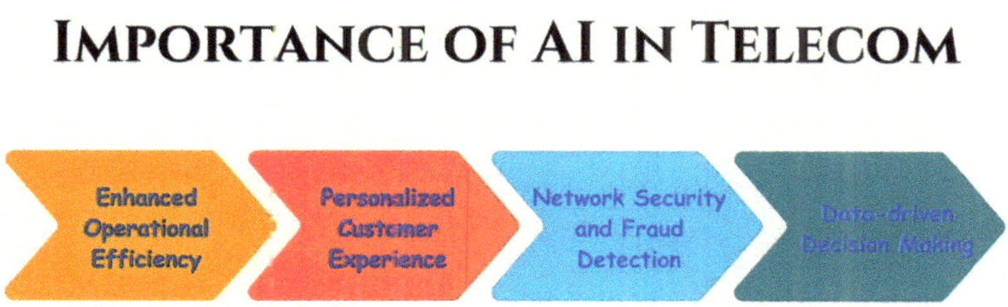

Figure 2.2 The importance of AI in telecom [2].

Artificial intelligence (AI) is transforming industries across the globe and telecommunications industry is no exception. The telecommunications industry has experienced a rapid transformation in recent years, and the advent of generative AI is set to bring about even more dramatic changes. It is at the forefront of technological innovation, and AI is playing a major role in this transformation. So many telcos are interested in AI because the opportunity to transform themselves is so vast [3].

This chapter examines the roles of AI and generative AI in the telecommunications industry. It begins with explaining artificial intelligence. It presents AI and generative AI in telecommunications. It highlights the benefits and challenges of AI in telecommunications. The last section concludes with comments.

2.2 WHAT IS ARTIFICIAL INTELLENCE?

The term "artificial intelligence" (AI) is an umbrella term John McCarthy, a computer scientist, coined in 1955 and defined as "the science and engineering of intelligent machines." It refers to the ability of a computer system to perform human tasks (such as thinking and learning) that usually can only be accomplished using human intelligence [4]. Typically, AI systems demonstrate at least some of the following human behaviors: planning, learning, reasoning, problem solving, knowledge representation, perception, speech recognition, decision-making, language translation, motion, manipulation, intelligence, and creativity.

The 10 U.S. Code § 2358 define artificial intelligence as [5]:

1. "Any artificial system that performs tasks under varying and unpredictable circumstances without significant human oversight, or that can learn from experience and improve performance when exposed to data sets.
2. An artificial system developed in computer software, physical hardware, or other context that solves tasks requiring human-like perception, cognition, planning, learning, communication, or physical action.
3. An artificial system designed to think or act like a human, including cognitive architectures and neural networks.
4. A set of techniques, including machine learning, that is designed to approximate a cognitive task.
5. An artificial system designed to act rationally, including an intelligent software agent or embodied robot that achieves

goals using perception, planning, reasoning, learning, communicating, decision making, and acting."

AI provides tools creating intelligent machines which can behave like humans, think like humans, and make decisions like humans. The main goals of artificial intelligence are [6]:

1. Replicate human intelligence
2. Solve knowledge-intensive tasks
3. Make an intelligent connection of perception and action
4. Build a machine which can perform tasks that requires human intelligence
5. Create some system which can exhibit intelligent behavior, learn new things by itself, demonstrate, explain, and can advise to its user.

AI is not a single technology but a range of computational models and algorithms. The concept of AI is an umbrella term that encompasses many different technologies. AI is not a single technology but a collection of techniques that enables computer systems to perform tasks that would otherwise require human intelligence. The major disciplines in AI include [7]:

- *Expert systems*
- *Fuzzy logic*
- *Neural networks*
- *Machine learning (ML)*
- *Deep learning*
- *Natural Language Processors (NLP)*
- *Robots*

These computer-based tools or technologies have been used to achieve AI's goals. Each AI tool has its own advantages. Using a combination of these models, rather than a single model, is recommended. Figure 2.3 shows a typical expert system, while

Figure 2.4 illustrates the AI tools. These tools are gaining momentum across every industry. Analytics can be considered a core AI capability.

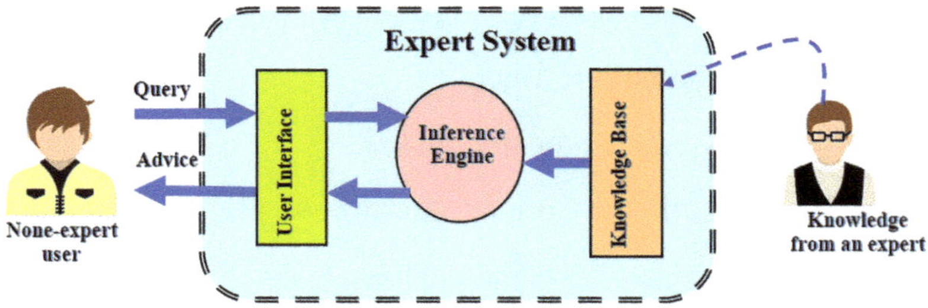

Figure 2.3 A typical expert system.

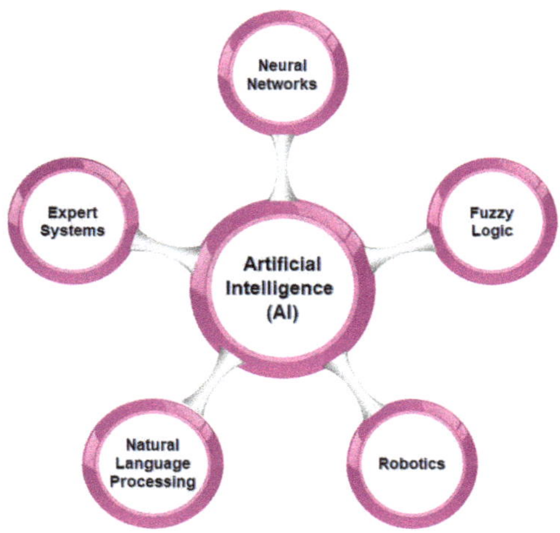

Figure 2.4 AI tools.

2.3 ARTIFICIAL INTELLIGENCE IN TELECOMMUNICATIONS

Almost every industry has been transformed by artificial intelligence, including the telecom sector. The telecommunications industry is evolving rapidly, and artificial intelligence (AI) is playing a pivotal role in shaping its future. The growth in the AI market size within this industry is directly linked to its value. AI is creating business value in terms of improved performance, higher efficiency, enhanced customer experience as well as creating new business models and use cases. By adopting AI solutions, you can help improve customer service, increase self-service offerings, empower your employees, and drive revenue growth. Artificial intelligence is being increasingly used in the telecommunications industry to analyze data, predict future trends and make informed decisions. Figure 2.5 shows the use cases of AI in telecommunications [8]. Here are essential applications of AI telecom [9]:

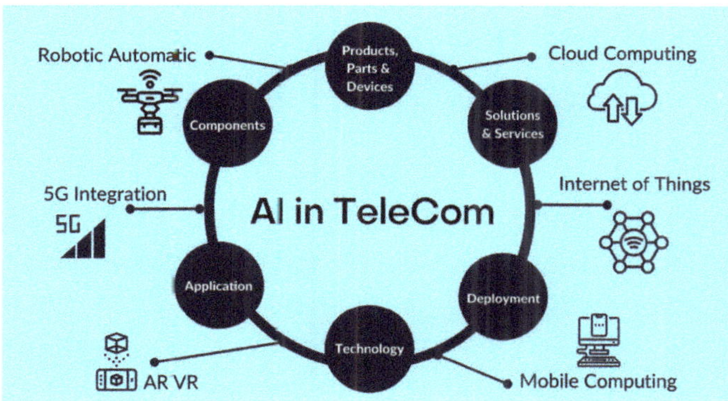

Figure 2.5 Use cases of AI in telecommunications [8].

- *Predictive Maintenance:* AI technology assists with predictive maintenance. This allows a Telco to predict future trends and patterns based on historical data to identify network and operational malfunctions. In this way, Telcos can predict hardware failures before they occur, allowing a Telco to be proactive in maintaining its equipment and improving overall network quality and customer experience. Combining the right technologies can enable them to shift to predictive maintenance, in which they leverage the vast stores of data that reflect how their infrastructure components are actually being used. Predicting failure rath than assuming it enables operators to maximize the life of each asset. This enables the telecom carrier to maximize network uptime, plan for CapEx and OpEx spending, and drive efficiency.

- *Fraud Mitigation:* Telecom fraud has been a challenge since the industry began. Today, AI-enabled tools let carriers stay ahead of malicious actors' evolving tactics. They also enable constant monitoring to identify when bots are using your network, prevent malicious actors from accessing personal customer information and other sensitive data, and prevent other unauthorized access. Telecoms are harnessing AI's powerful analytical capabilities to combat fraud. AI and machine learning algorithms can detect anomalies in real-time, effectively reducing telecom-related fraudulent activities, such as unauthorized network access and fake profiles. The system can automatically block access to the fraudster as soon as suspicious activity is detected, minimizing the damage.

- *Customer Service:* Customer satisfaction plays a vital role in any business to prosper. Customer service and related issues are a constant struggle. This involes combining machine learning (ML) and AI with natural language processing (NLP). This requires the carrier to determine the ideal balance of human skills and this powerful combination can free up human workers to take on more complex and valuable tasks. Beyond just chatbots and customer service assistants, a strong customer data platform (CDP) enables marketers to create customer journey maps and update them in real time. Telecom companies use artificial intelligence and machine learning to improve customer service through chatbots and virtual assistants. Chatbots assist in analyzing requests, routing, and escalating customer inquiries. The AI-driven assistants, employing natural language processing, swiftly address consumer questions. Telecom virtual assistant can handle most inquiries, from billing to technical issues, ensuring comprehensive support. Figure 2.6 shows a typical telco customer [10].

Figure 2.6 A typical telco customer [10].

- *Network Optimization:* One of the most impactful applications of AI in telecommunications is network optimization. Traditional network management systems often struggle with the growing complexity and data volumes. AI algorithms, however, can process vast amounts of data in real-time, enabling more efficient network management and resource allocation. The proactive approach not only improves user experience but also reduces operational costs. As 5G networks began to roll out in 2019, AI became essential for helping CSPs (communications service providers) build self-optimizing networks (SONs) to support predicted growth. These networks allow operators to automatically optimize network quality based on traffic information by region and time zone.

- *Cybersecurity:* Telecommunications networks are prime targets for cyberattacks and fraudulent activities due to their massive scale and critical nature. AI is proving to be an invaluable asset in protecting these networks from malicious actors. Static rules and signatures are the foundation of traditional security solutions, which can easily become out-of-date and unable to handle sophisticated attacks that are constantly changing and aimed at communications service provider (CSP) networks. AI systems are able to adjust to the ever-evolving threat landscape, identifying abnormalities as harmful on their own and providing context to assist human specialists.

2.4 GENERATIVE AI IN TELECOMMUNICATIONS

The telecom industry can achieve new levels of profitability with generative AI. The role of AI in the telecommunications industry is growing, especially with the emergence of generative AI. Artificial intelligence (AI) and generative AI (GenAI) are changing how we approach mobile and connectivity requirements. Generative AI has taken the world by storm, and exciting use cases are already emerging in the telecom industry. It is an advanced technology with the potential to transform telecom companies' operations, customer interactions, and service offerings. Figure 2.7 shows some generative AI use cases in the telecom industry [11]. Some of these use cases are explained as follows [12]:

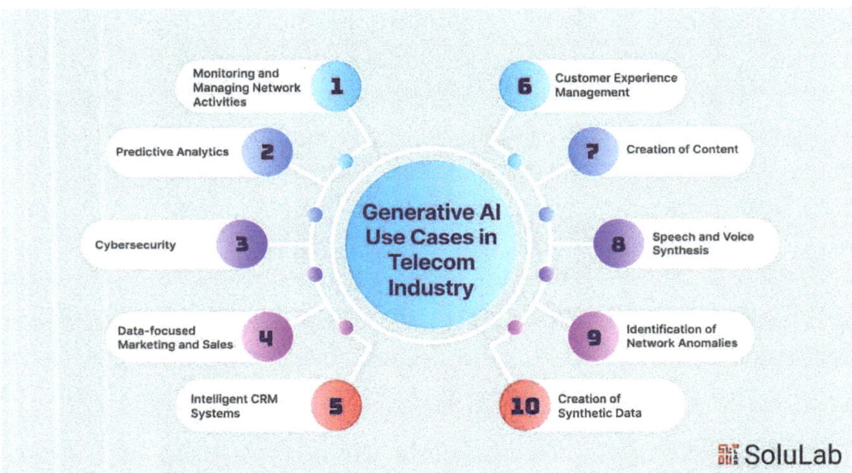

Figure 2.7 Some generative AI use cases in the telecom industry [11].

• *Network Management:* Generative AI connects multiple complex AI/ML models used across network planning and operations. They understand network behaviors and create action plans in areas like network capacity planning and performance. Technology can train models with customer experience and sentiment data to build better prediction capabilities, significantly enhancing privacy, factuality, and relevance while protecting intellectual property. For example, AI can automate network management for 4G and 5G systems. Generative AI for telecom enhances performance by producing summaries and recommendations for network orchestration or coding.

- *Customer Service:* Generative AI in telecom can analyze a customer's usage patterns and interactions with the provider's website and app to recommend personalized plans and add-ons. If customers exceed their data limit, AI can suggest a plan with more data and add-ons like international calling or unlimited texting. AI chatbots and virtual assistants provide 24/7 support, handling customer queries efficiently and generating human-like responses. AI chatbots can address various issues, from billing inquiries to technical support, and escalate complex problems to human agents when necessary.

- *Marketing:* Data-driven marketing and sales is a critical use case of AI in the telecom industry, enabling companies to harness the power of customer data for strategic decision-making. Generative AI can create tailored content for different audiences and channels, such as blog posts, social media, landing pages, and email campaigns. Telecom operators can use AI to generate titles, summaries, keywords, or captions based on audience interests and actions. It can also classify and segment customers based on attributes, desires, preferences, and actions, using data from web analytics, CRM platforms, and social media, from customer interactions, and target customers using micro-segmentation derived from call insights.

- *Fraud Detection:* Since telecom networks handle vast amounts of sensitive data, security is a top concern. Generative AI in telecom can spot and counter security threats by detecting anomalies in network traffic in real time that indicate possible cyberattacks or unauthorized access. Telecommunications providers use Generative AI to protect against fraudulent activities by analyzing data, including call records and online transactions. It also identifies patterns of SIM card swapping or suspicious login locations, helping telecom companies swiftly detect and mitigate fraud.

- *Predictive Maintenance:* Predictive maintenance is another area where generative AI is having a profound effect. Generative

AI for telecom analyzes vast amounts of data to predict equipment failures and network issues before they occur. The AI promptly raises alarms when anomalies or deviations arise, such as unexpected traffic spikes or equipment malfunctions. This proactive monitoring allows telecom operators to address potential issues through automated responses swiftly. Telecommunication companies can be enabled to do proactive maintenance scheduling during off-peak hours, minimizing disruptions and ensuring consistent connectivity.

- *Billing:* AI-based billing is one of the promising AI use cases in telecommunications. Generative AI refines billing inquiries by providing precise solutions and insights. AI algorithms can analyze usage data to calculate bills accurately, eliminating errors, and enhancing customer trust. For example, AI can provide detailed billing explanations, helping customers understand their charges and reducing billing-related disputes. By using AI-based billing, companies can also provide personalized bill explanations to customers, improving transparency and trust.

- *Data Generation:* Synthetic data generation for testing, training, and research involves creating realistic network traffic patterns to test and strengthen security systems against potential cyber-attacks. By simulating various attack scenarios, AI helps in identifying vulnerabilities and fortifying defenses, ensuring robust network security. For example, Generative AI for telecom can expedite the resolution of fiber cut incidents, a common issue in telecom.

2.5 BENEFITS

Telcos are using AI to discover valuable information, enhance customer satisfaction, streamline operations, and boost revenue. Adopting this technology dramatically enhances Telcos' ability to respond to potential threats quickly, protecting systems and data before detrimental consequences or losses can occur. The technology is used to answer several industry challenges. The benefits of AI can be seen across many areas of their business, from network optimization to revenue growth. Communication service providers (CSPs) that implement artificial intelligence in their telecom operations are twice as likely to manage bigger operations efficiently as their manual counterparts. Other benefits include [13]:

- *Automation:* Repetitive tasks, often considered monotonous by humans, find a perfect match in AI algorithms, which never experience boredom, fatigue, or the need for rest. The AI technology automates lengthy procedures, such as billing, order management, and data processing, benefiting telcos in several ways. By leveraging these technologies for automation, telcos can enhance operational efficiency, optimize cost management, minimize resource requirements, and mitigate the risk of human error, ultimately resulting in reduced operational expenses.

- *Data Analysis:* Data is one of Telcos' most significant assets, but your data is only as good as the insights you get from it. With so much data coming from different data streams, managing, interpreting, and extracting relevant and valuable insights is often time-consuming. AI simplifies extensive data analysis by automating the data process, including data preparation, visualization, and predictive modelling.

- *Predictive Analytics:* AI is used for predictive analytics. These analytics allow telcos to extract valuable insights from massive amounts of data. AI-driven predictive analytics are helping telecoms provide better services by utilizing data, sophisticated algorithms, and machine learning techniques to predict future results based on historical data. By using this data to examine customer behavior, life cycles, purchasing decisions, etc., a telco can gain insight into what customers are looking for and plan products and solutions to meet these needs and demands.

- *Improved Network Performance:* One of the most important ways that AI is being used in the telecom industry is to improve network performance. AI can be used to analyze data from network sensors to identify potential problems before they occur. This allows telecom providers to take proactive steps to fix problems and prevent outages. AI is being used to analyze data from network sensors to identify potential problems before they occur. This allows telecom providers to take proactive steps to fix problems and prevent outages.

- *Improved Customer Experience:* Telcos see improved customer experience and optimized network operations as the top two benefits of integrating AI into their operations. The insights to better meet the expectations of those customers with more tailored, specific services.

- *Swift Decision-making:* AI can make decisions in the blink of an eye, processing thousands of data items in mere fractions of a second. For example, AI systems can monitor thousands of credit card transactions in real-time, swiftly identifying and blocking potentially fraudulent ones.

- *Expertise:* While humans invest years in acquiring expertise, AI offers a way to leverage rare and extensive knowledge more efficiently. Expertise can be distilled into a model, making it applicable across a broader spectrum. Expert knowledge can be extended to scenarios where it was previously lacking, allowing human experts to allocate their time more efficiently. Today, AI has become far cheaper and faster to train for a much broader range of capabilities

- *Robotic Process Automation (RPA):* Many enterprises have vast numbers of customers engaged in millions of daily transactions, each susceptible to human error. Robotic Process Automation (RPA) is a business process automation technology based on AI. RPA has always been the number one choice for all digital transformation projects. If implemented correctly, it will deliver tangible value from day one by reducing document processing times and accelerating business flows. With AI applied to RPA, the performance-boosting effect is even more profound, allowing for anomaly detection and (semi-)automatic error correction.

- *Personalized Assistance:* Telecom companies can now provide customers with personalized assistance and answers, all in a friendly, human-like way. In the not-so-distant future, we might bid farewell to traditional human customer service agents as virtual assistants and chatbots take center stage. This will cut costs, make customers happier, and keep them coming back for more. AI-driven assistants analyze client data, offering personalized recommendations. They also create proactive, transformative customer interactions, fostering loyalty, and driving revenue growth.

- *Revenue Growth:* AI can take a lot of different types of data, like what devices people are using, and make sense of it. This can help telecom companies grow their business by ensuring they have enough subscribers and making money from each one. Also, AI can help telecom companies offer their products to customers at the right time and place.

- *Energy Efficiency:* In view of the escalating cost of electricity, it has been natural for operators to employ AI to optimize power consumption in their networks. Telefónica Spain was the first operator worldwide to test Deep Sleep Mode energy saving functionality.

- *Anomaly Detection:* Anomaly detection identifies unusual patterns that could indicate threats or system failures, enabling quick mitigation efforts. These systems monitor network traffic, performance data, and security logs for patterns or activities that deviate from the norm. These activities include unusual network traffic that indicates a cybersecurity threat to unexpected drops in performance. Figure 2.8 shows some of these benefits [14].

Figure 2.8 Some benefits of AI in telecom [14].

2.6 CHALLENGES

Several challenges continue to hinder the widespread adoption of AI and generative AI in the industry. There are no shortcuts; successful implementation of AI and generative AI requires strategic investment in infrastructure, skills, and change management. Telecommunications companies have amassed vast troves of data from their extensive customer bases over the years. However, harnessing the full potential of this data has been a challenge. Other challenges include the following [13]:

- *Security:* Security has always been a major concern for telecom companies as they hold a vast amount of customers' data including their financial details. AI involves processing vast amounts of sensitive customer data, which must be protected against breaches and misuse. In today's world, the prevalence of online criminal activity, particularly fraud, poses a substantial risk to telcos. Advanced algorithms in AI technology play a crucial role in identifying and anticipating any unusual activities occurring within the network. This allows telcos to swiftly identify and respond to any security or data breaches.
- *Privacy:* In addition to security, telecom companies must navigate complex regulations related to data privacy, such as the General Data Protection Regulation in Europe and the California Consumer Privacy Act in the United States. Non-compliance can result in serious fines and legal repercussions.
- *Skill Shortage:* A lack of skills or finding the appropriate skilled labor is the top challenge for AI and generative AI. In other words, the successful implementation of generative AI faces a challenge due to the requirement for more proficient AI professionals. To overcome this obstacle, it is crucial to invest in training programs and recruit data science experts, as this will aid in bridging the skill gap within the industry.
- T*echnical Challenges:* Other barriers include a mix of technical challenges—including poor technology infrastructure, a lack of data, and data privacy concerns—

and budget constraints.

- *Trust:* Building trust in AI-based systems is crucial because AI is frequently used to support human decision-making while avoiding the development of excessive reliance on technology. AI systems run the risk of being underutilized in the absence of trust. This could then result in a sizable loss of potential value creation for both businesses and society at large. Similarly, excessive AI use and reliance may be harmful. To realize the full potential of AI, trust needs to be established in the development, deployment, and use of AI. This is why it is critical for us to build human trust in AI. Trustworthiness is a prerequisite for AI and it must be built into the system by design.
- *Ethics:* Ethics are central to the success of AI – and how AI can reduce security and compliance issues, detect threats like cyberattacks or fraud and improve customer data protection, driving a new era of security in telecom networks.
- *Integration Complexity:* Integrating generative AI into existing systems can be complex. Telecom companies often operate on legacy systems unsuitable for advanced AI technologies. These systems, often built decades ago, lack the flexibility and scalability required to support advanced AI technologies. The incompatibility can lead to delays in deployment, increased operational costs, and potential service disruptions.
- *Data Quality:* AI is only as good as the data it is built on, which means you need to ensure that the data you receive is credible and complete. Telecom companies generate a plethora of data across various departments and legacy systems, including customer information, network performance metrics, and service usage statistics. This data often resides in silos, making it difficult to consolidate and analyze effectively. Ensuring data quality is another critical issue; poor data quality can lead to inaccurate AI predictions and unreliable outputs. Do not let imperfect data get in the

way of making progress. Use the data you have available now while broader data modernization initiatives are being stood up.

- *High Complexity:* Implementing generative AI in telecom involves significant complexity and risks. The development and deployment of AI models need precise calibration, continuous monitoring, and regular updates to adapt to changing network conditions and user behaviors. Any errors or inaccuracies in the generative AI models can result in service disruptions and financial losses.
- *Poor Network Management:* Global traffic and the need for more network equipment are growing dramatically, resulting in more complex and costly network management.

2.7 CONCLUSION

Integrating AI in the telecoms industry is a continuous and evolving process with the potential to enhance efficiency, improve customer experiences, and drive innovation across various aspects of telecommunications. To meet new customer demands, empower employees, reduce total cost to serve and grow revenue, telecommunications providers should embrace and adopt AI solutions. Each company is at a different stage of adopting AI and GenAI solutions into its everyday operations.

The influence of artificial intelligence in telecom is becoming more prevalent as the industry continuously explores the potential of AI. As big data tools and applications become more available and sophisticated, the future of AI in the telecom industry will continue to develop. AI technology is here to stay and will continue to significantly impact the telecommunications industry. More information about the role of AI in the telecommunications can be found in the books [15,16] and the following related journal:

- *Energy and AI*
- *The AI Journal*

REFERENCE

[1] "Modi 3.0: DPDP, sat spectrum norms, electronics PLI boost on cards,"

https://www.business-standard.com/industry/news/notifying-rules-for-dpdp-telecom-acts-to-keep-it-telco-ministers-busy-124060900271_1.html

[2] "Top 10 use cases of AI in telecom for 2024," June 2024,

https://www.appsdevpro.com/blog/ai-in-telecom-top-use-cases/

[3] M. N. O. Sadiku, O. D. Olaleye, and J. O. Sadiku, "Artificial intelligence in telecommunications," *International Journal of Trend in Research and Development*, vol. 11, no. 6, November-December 2024, pp. 127-133.

[4] M. N. O. Sadiku, "Artificial intelligence," IEEE Potentials, May 1989, pp. 35-39.

[5] "Artificial intelligence (AI),"

https://www.law.cornell.edu/wex/artificial_intelligence_(ai)

[6] "Artificial intelligence tutorial,"

https://www.javatpoint.com/artificial-intelligence-tutorial

[7] D. Quinby, "Artificial intelligence and the future of travel," May 2017,

https://www.phocuswright.com/Travel-Research/Research-Updates/2017/Artificial-Intelligence-and-the-Future-of-Travel

[8] "AI & telecommunications: The ultimate revelation by top 10 experts," September 2021,

https://www.telecloudvoip.com/ai-in-telecom-revelation-industry-experts/

[9] "Eight AI use cases in the telecom industry,"

https://www.sandtech.com/insight/eight-ai-use-cases-in-the-telecom-industry/

[10] "Transforming telecom with next-gen AI," Unknown Source.

[11] "The impact of generative ai on the telecommunication industry,"

https://www.solulab.com/generative-ai-in-telecom-industry/

[12] "Practical use cases of generative AI in telecom," July 2024,

https://www.linkedin.com/pulse/practical-use-cases-generative-ai-telecom-n-ix-6s2gf/

[13] H. V. Heerdan, "The impact of AI in telecommunications," December 2023,

https://telecoms.adaptit.tech/blog/the-impact-of-ai-in-telecommunications/

[14] "AI in telecom – Exploring the key business benefits, use cases, examples and challenges," May 2024,

https://appinventiv.com/blog/ai-in-telecom/

[15] Y. Kambayashi et al. (eds.), *Artificial Intelligence for Communications and Networks: 4th EAI International Conference, AICON 2022, Hiroshima, Japan, November 30 - December 1, ... and Telecommunications Engineering)*. Springer, 2023.

[16] N. Bouguila and Y. Laalaoui, *Artificial Intelligence Applications in Information and Communication Technologies*. Springer, 2015.

CHAPTER 3
DRONES IN TELECOMMUNCATIONS

"If the government regulates against use of drones or stem cells or artificial intelligence, all that means is that the work and the research leave the borders of that country and go someplace else."

—Peter Diamandis

3.1 INTRODUCTION

Drones have evolved in the past few years to become exceptionally versatile across practically every industry. From improving efficiency to reducing risks, the various commercial uses for drones make them indispensable tools in many sectors. Drones can quickly identify potential issues, such as damaged equipment or vegetation encroachment, allowing for more efficient maintenance and reducing the risk of service disruptions.

Drone technology has exciting growth potential across many industries. One such industry is telecom. Telecommunications industry reaches into every corner of our economies, societies, and private lives, and it is one of the greatest drivers of economic growth and human equality. It encompasses a wide array of services and technologies, including telephone services, Internet access, and broadcasting. Telecom operators increasingly rely on drone services to inspect and maintain communication towers and other critical infrastructure. Every telco should investigate how drones can lower their cost, speed up their deliverables, reduce risk, and provide additional revenue streams.

The integration of drones with other technologies, such as AI and IoT, will further enhance their capabilities and create new opportunities for growth and optimization. Telecom operators have started drone cell tower inspections to assess network coverage and performance, map tower structures, and provide real-time data to repair crews.

Drone applications enable various types of companies to transform their operations and gain efficiency. One of the sectors that has seen a radical transformation due drone adoption is telecommunications. Drone technology is changing the face of telecommunications quickly and its first real application in telcos is tower inspections.They are becoming increasingly popular in the telecom industry as they provide a more efficient and cost-effective way for technicians to conduct tower inspections and surveys.The use of drones in the tower industry is growing quickly for network operators, tower engineers, tower owners, and tower climbers [1].

This chapter examines the applications of drones in the telecommunications industry. It begins with describing what a drone is. It covers telecommunications drones. It presents some applications of drones in the telecommunications industry. It highlights some benefits and challenges of drones in telecommunications. The last section concludes with comments.

3.2 WHAT IS A DRONE?

The FAA defines drones, also known as unmanned aerial vehicles (UAVs), as any aircraft system without a flight crew onboard. Drones include flying, floating, and other devices, including unmanned aerial vehicles (UAVs), that can fly independently along set routes using an onboard computer or follow commands transmitted remotely by a pilot on the ground. A typical drone is shown in Figure 3.1 [2]. A drone is usually controlled remotely by a human pilot on the ground, as typically shown in Figure 3.2 [3].

Figure 3.1 A typical drone [2].

Figure 3.2 A drone is usually controlled by operators on the ground [3].

Drones can range in size from large military drones to smaller drones. Drones, previously used for military purposes, have started to be used for civilian purposes since the 2000s. Since then, drones have continued to be used in intelligence, aerial surveillance, search and rescue, reconnaissance, and offensive missions as part of the military Internet of things (IoT). Today, drones are used for different purposes such as aerial photography, surveillance, agriculture, entertainment, healthcare, transportation, law enforcement, etc. Commercial drones have come a long way in the last decade.

Drones work much like other modes of air transportation, such as helicopters and airplanes. When the engine is turned on, it starts up, and the propellers rotate to enable flight. The motors spin the propellers and the propellers push against the air molecules downward, which pulls the drone upwards. Once the drone is flying, it is able to move forward, back, left, and right by spinning each of the propellers at a different speed. Then, the pilot uses the remote control to direct its flight from the ground [4].

Drone laws exist to ensure a high level of safety in the skies, especially near sensitive areas like airports. They also aim to address privacy concerns that arise when camera drones fly in residential areas. These include the requirement to keep your drone within sight at all times when airborne. In the United States, drones weighing less than 250g are exempt from registration with civil aviation authorities. If your drone exceeds 250g in weight, you will also require a Flyer ID, which requires passing a test [5]. It is necessary to register as an operator, be trained as a pilot, and have civil liability insurance, in addition to complying with various flight regulations, and those of the places where their use is permitted.

Most drones have a limited payload, usually under 11 pounds. Drones are classified according to their size. Here are the different drone types:

- Nano Drone: 80-100 mm
- Micro Drone: 100-150 mm
- Small Drone: 150-250 mm
- Medium Drone: 250-400 mm
- Large Drone: 400+ mm

One of the emerging trends in drone use for factories is the utilization of LiDAR technology. LiDAR stands for Light Detection and Ranging. This technology provides accurate depth information essential for understanding the three-dimensional structure of the environment. LiDAR sensors emit laser beams to measure distances to objects, creating high-resolution 3D maps of the surrounding terrain and objects. The ability to capture detailed data through LiDAR technology has opened up opportunities for better predictive maintenance, reduction in inspection times, and overall cost savings [6].

3.3 TELECOMM DRONES

The telecommunications industry is a critical sector that facilitates the transmission of information across distances through electronic means. This industry relies on complex infrastructure such as satellites, fiber optics, and cellular networks to provide seamless communication [7]. The telecom sector accounts for 8.5 per cent of the GDP of US.

Drones have already revolutionized several industries such as business, entertainment, manufacturing, military operations, security, search and rescue, and inspections. The telecommunications industry is one of the first industries to benefit from the use of drones. Payload drones in the telecommunications industry are sometimes called cargo drones, delivery drones, transport drones, load-carrying drones, freight drones, logistics drones, utility drones, aerial transport vehicles, load-bearing UAVs (unmanned aerial vehicles), and heavy-lift drones. Telecom operators are considering whether to build their drone solutions or embrace third-party drone services for telecom drones, such as cell tower inspections and telecom tower inspections. A typical drone used for telecommunications is shown in Figure 3.3 [8].

Figure 3.3 A typical drone used for telecommunications [8].

Telecommunications inspections have become much more complex. Drone adoption is happening exponentially as the cost of drones decreases each year and the technology continues to evolve to provide autonomous flight, collision avoidance, auto landing, and communications. The uses of drone technology in the telecom sector include [9].

- Inspections of towers and tower-based equipment
- Line of sight testing
- Signal strength and coverage testing
- Maintenance and monitoring
- Leveraging drones in low-latency IoT

3.4 APPLICATIONS

Drones are increasingly utilized in the telecommunications industry. They are revolutionizing the way we monitor and safeguard our energy infrastructure. Common applications include the following [8,10].

- *Infrastructure Inspection:* Drones have become an indispensable tool for tower inspections and maintenance in the telecommunications industry. Infrastructure inspection and maintenance using commercial drones have become increasingly popular due to their ability to access hard-to-reach locations, reduce costs, and improve safety. Tower inspections is one area where drones excel and they deliver high performance inspections. It is no longer necessary for employees spend hours setting up cranes to mount a camera in the air. Drones have proven to be invaluable tools for inspecting and maintaining telecommunications infrastructure and power lines. Drones can detect structural issues or potential hazards by inspecting structures such as bridges and buildings, enabling timely and targeted repairs. Equipped with high-resolution cameras and other sensors, drones can safely fly close to transmission lines, towers, and antennas, capturing detailed images and data for analysis. Figure 3.4 shows using a drone for tower inspection [11].

Figure 3.4 Using a drone for tower inspection [11].

- *Rescue Operations:* Drones have become indispensable tools in search and rescue operations, improving the speed and efficiency with which missing persons can be located. In situations where time is of the essence, this rapid response capability can mean the difference between life and death. By reducing the time and resources required to conduct search and rescue operations, drones improve the likelihood of successful rescues.

- *Surveying and Mapping:* Drones play a critical role in surveying and mapping for the telecommunications industry. They can rapidly generate accurate 3D models and high-resolution orthomosaic maps of the terrain. Drones can cover large areas quickly, making them ideal for collecting data in remote or hard-to-reach locations.

- *Network Expansion:* As the demand for better and faster connectivity continues to grow, the telecommunications industry must constantly expand and improve its networks. Drones help achieve this by assisting in the installation of new equipment and monitoring the progress of ongoing projects. Drones also enable real-time monitoring of construction sites, ensuring that projects stay on schedule and meet quality standards.

- *Signal Strength Testing:* Drones equipped with specialized antennas and signal testing equipment can perform aerial signal strength testing. This helps engineers fine-tune the positioning and alignment of antennas, improving network performance and coverage.

- *Network Planning:* Drones play a significant role in network planning by providing accurate and up-to-date data on terrain, infrastructure, and signal strength. This information helps engineers design and optimize networks for maximum coverage and performance.

- *Emergency Response:* Drones have been utilized to improve emergency response in the telecommunications industry. In the aftermath of a natural disaster, such as a hurricane or earthquake, drones can be deployed to assess the damage to telecommunications infrastructure and identify areas where repairs are needed. This rapid assessment allows companies to restore service more quickly.

3.5 BENEFITS

The commercial use of drones is transforming various industries and providing significant benefits, from increased efficiency and cost savings to improved safety and reduced environmental impact. Drones reduce the need for manual labor and lower the risks associated with working at height or in confined spaces. A drone is unmanned, so if there is an accident in the air, only the equipment is harmed. As the benefits of drone technology in telecommunications become more widely recognized, we can expect to see greater adoption across the industry. Other benefits of using drones in telecommunications include the following [10,12,13]:

- *Connectivity:* The telecommunications industry plays a vital role in keeping us connected. As the telecom industry grows and expands, drones will be crucial in supporting its infrastructure and connectivity needs. Drones are being used to connect inaccessible areas, providing connectivity across different parts of the world. Adopting drone technology in the telecom industry is essential for developing and deploying 5G connectivity.

- *Cost Savings:* Drone technology can help reduce the costs of tower inspections and surveying. Drones offer significant cost savings in the telecommunications industry by reducing the need for manual labor and expensive equipment. They can minimize downtime by identifying issues before they cause network outages, reducing maintenance costs. By using drone technology, companies can streamline the planning process, and reduce costs,

- *Improved Safety:* Traditional methods require a team of experienced linemen to visually inspect all lines in an area to evaluate the lines for any damage or potential problems. Utilities are increasingly turning to drones for the inspection of powerlines and pipelines in hard-to-reach areas or in difficult terrain. Using drones in telecommunications greatly enhances worker safety. Climbing towers for inspections and maintenance can be dangerous, and accidents can result in injuries or even fatalities. Drones eliminate the need for technicians to climb towers in many cases, as typically shown in Figure 3.5 [14]. This reduces the risk of accidents and promoting a safer work environment. Drones equipped with cameras and sensors can provide additional security for telecom infrastructure, helping to detect and deter potential vandalism or theft.

Figure 3.5 Drones eliminate the need for technicians to climb towers [14].

- *Increased Efficiency:* Accurate cell tower inspections in a matter of minutes using drones. Drones can perform tasks more quickly and accurately than humans, increasing efficiency in the telecommunications industry. They can inspect towers, map terrain, and test signal strength in a fraction of the time it would take a team of technicians.

- *Better Data Collection:* Companies large and small seek ways to collect data faster and lower risk and cost. Drones equipped with advanced sensors and cameras can collect valuable data that can be used to optimize networks, monitor infrastructure, and plan future expansions. The data collected provides companies with the information they need to make data-driven decisions and improve their operations.

- *Real-time Data:* Drones can provide real-time data and analytics and help decrease tower downtime. Real-time aerial data analytics for tower inspection allows operators to collect, process, and analyze data in real-time. Drones for tower inspection can capture high-resolution images and data from difficult-to-reach areas and use real-time data analytics and artificial intelligence (AI) to analyze structural data and identify problem areas. Telcos can use real-time data from drones to improve wireless network performance, identifying and addressing problem areas, and quickly resolving issues to improve the overall quality of service.

- *Access to Remote Areas*: Drones can reach areas that are difficult to access, allowing telecom providers to expand their geographic presence. They can quickly access hard-to-reach areas, such as cell towers or transmission lines, facilitating timely repairs and minimizing the need for manual inspections.

- *Faster Inspection:* Drones have become an indispensable tool for tower inspections and maintenance in the telecommunications industry. They can be used to conduct inspections more frequently, allowing for early detection of small issues. They can cover a larger area in a shorter amount of time than traditional inspection methods. Drones can gather information more efficiently than human inspectors, and the digital data can help employees make better decisions.

- *Reduced Risk:* Drones can speed up message and response times, which can reduce the risk of damage to pipelines. They significantly improve safety, save time, and minimize downtime.

- *Surveying:* Drones play a critical role in surveying and mapping for the telecommunications industry. They can rapidly generate accurate 3D models and high-resolution orthomosaic maps of the terrain, helping engineers identify the best locations for new towers or network expansions. Drones can cover large areas quickly, making them ideal for collecting data in remote or hard-to-reach locations. This accelerates the planning process and ensures optimal network coverage.

- *Network Expansion:* As the demand for better and faster connectivity continues to grow, the telecommunications industry must constantly expand and improve its networks. Drones help achieve this by assisting in the installation of new equipment and monitoring the progress of ongoing projects. They can transport small payloads, such as antennas and cables, to hard-to-reach locations, speeding up the deployment process.

- *Timely Maintenance:* Drones are used for timely maintenance and repairs in the telecommunications industry. Drones equipped with remote monitoring capabilities allow for efficient and accurate inspections of telecommunication infrastructure. By employing predictive analytics, drones can identify potential issues before they lead to costly downtime or service interruptions. Through real-time data collection and analysis, drones can detect anomalies in equipment performance or identify areas requiring maintenance.

- *Environmental Benefits:* By incorporating drones into operations, the telecommunications industry can reap various environmental benefits. Drones have the potential to revolutionize wildlife monitoring and precision agriculture, leading to a more sustainable future. By identifying areas of crop stress, drones can enable targeted interventions, reducing the overall environmental impact of agriculture. Incorporating drones into operations can thus contribute to the preservation of biodiversity and the promotion of sustainable farming practices.

- *Competitive Advantage:* Incorporating drones into operations in the telecommunications industry can provide a significant competitive advantage. Drones have the potential to revolutionize the way businesses in this sector operate, offering strategic differentiation and the opportunity for market dominance.

Embracing drone technology positions businesses in the telecommunications industry as innovative and forward-thinking, helping them gain a competitive edge in the market.

3.6 CHALLENGES

There are several challenges facing the use of drones in telecommunications. The main challenge is the need for skilled drone operators with experience in the telecom and drone industries. Access to high ground in rural areas is another challenge. Drones will not replace engineers nor tower climbers, but they are enabling them to work faster and more efficiently. Drone technology also comes with privacy concerns, regulatory challenges, and potential security risks. Other challenges facing drones in telecommunications include the following [10,15]:

- *Privacy:* Privacy is perhaps the most controversial matter when it comes to malicious drone use. Drones used in the telecommunications industry often collect large amounts of data, including images and videos of infrastructure and terrain. Operators must comply with privacy regulations and ensure that any collected data is stored and used in accordance with applicable laws. Drone operators can help ensure the safe and responsible integration of drones into the telecommunications industry.

- *Regulation:* As drones become more common in the telecommunications industry, it is essential to be aware of the regulations that govern their use. These regulations are in place to ensure the safety of both drone operators and the general public. Since drone usage is still a relatively new, several regulatory issues and grey areas must be addressed to integrate drones into the telecom industry successfully. As drones become more common in the telecommunications industry, it is essential to be aware of the regulations that govern their use. These regulations are in place to ensure the safety of both drone operators and the general public. Drones used for commercial purposes, such as in the telecommunications industry, must be registered with the relevant aviation authority. As drones become more integrated into the telecommunications industry, regulations will need to adapt to accommodate these new applications.

• *Airspace Restrictions:* Drone operations in the telecommunications industry must adhere to airspace restrictions and follow any applicable rules for flying near airports, heliports, or other sensitive areas.

• *Operator Licensing:* Telecom operators can carry out drone flights to execute necessary tasks, such as collecting images and video footage. Drone operation requires skilled pilots and regulatory compliance, which can be challenging for telecom operators. In the US, Canada, and other parts of the world, all drone operators must be duly licensed and maintain their piloting education to satisfy the rigorous criteria. Drones must adhere to specific aviation and privacy regulations.

• *Hatred for Drones:* Drone pilots are probably familiar with the experience of drawing small crowds whenever they are out flying drones. However, the opposite of this experience is also not uncommon, people demanding that these drones stay away from them. This is a story that gets told over and over, a person shoots down a drone that is flying near or above their property because of concerns over their safety or privacy, or simply because they believe that the drone should not be flying over their private property.

• *Drones for Evil Means:* While drones have proven essential for business and humanitarian operations. it is not impossible for drones to be used for evil means. Just like any technology, drones can be either good or bad depending on how people use them.

• *Expensive Technology:* As with any technology that pushes the boundary of what is possible, many of these advanced drones remain firmly beyond the budget of most people or organizations. Buying advanced equipment can provide a competitive edge, but these are certainly too expensive for a drone pilot who is still early in their career. As the commercial drone industry becomes more competitive, it becomes apparent that a drone pilot needs to have a somewhat expensive drone to even be noticed by clients. Add to that the costs of certification and training and you are really going to need significant investment to even get started in a professional drone career.

- *Skill Shortage:* There may be a lack of skilled drone engineers to set up and operate drone systems.

- *Communication:* Drones can face challenges with communication such as channel conditions, interference, antenna orientation, intermodulation distortion, spectrum availability, and network congestion. For example, when multiple drones and terminals share a wireless network, it can cause congestion. Signal Interference: Signal interference can limit the effectiveness of drones.

- *Logistics:* Drone delivery could pose logistical issues, esecially in densely populated areas. The technology would need to be capable of maneuvering around buildings and wildlife like birds. Also, deliveries to specific apartments would be difficult without human intervention. So you can get almost anywhere fairly quickly by truck or car. That makes it much harder for drones because the drones now have to compete against something that is been fine-tuned and optimized.

- *Vulnerability to Hacking:* Hackers can attack a drone's central control system and gain control of the drone.

- 3.7 CONCLUSION

Telecom operators should develop a tailored strategy and implementation road map for commercial drone applications. As drones in telecommunications industry become increasingly prevalent, they are transforming the way companies approach tower inspections, network expansions, and emergency response. As drone technology continues to advance, we can expect greater integration with other technologies, such as artificial intelligence (AI) and the Internet of things (IoT). The widespread deployment of 5G bandwidth will also improve drone performance. The bandwidth of 5G provides drones the ability to upload cloud imaging in real time. We can also expect to see even more innovative applications in the telecommunications industry.

The future of telecommunications is definitely wireless. The future of telecom drones looks bright and promising as drone technology continues to evolve and improve. Future advancements in drone technology could lead to improved capabilities and new applications in the telecommunications industry.

REFERENCE

[1] M. N. O. Sadiku, O. D. Olaleye, and J. O. Sadiku, "Drones in telecommunications," International Journal of Trend in Research and Development, vol. 11, no. 5, September 2024, pp. 102-105.

[2] D. Daly, "Tethered drones: A critical component of tactical communications," September 2021,

https://elistair.com/resources/military-drones/tethered-drones-a-critical-component-of-tactical-communications/

[3] "Revolutionize your inspections with WISPR American-made drones,"

https://wisprsystems.com/drones-for-inspection/?gad_source=1&gclid=EAIaIQobChMI9LvR9KyniAMV4ZHCCB-3JSygnEAAYASAAEgKFyPD_BwE

[4] "How drones work and how to fly them," May 2024,

https://dronelaunchacademy.com/resources/how-do-drones-work/

[5] "What are the main applications of drones?" June 2024,

https://www.jouav.com/blog/applications-of-drones.html

[6] "Drones in manufacturing: A game-changer for industry,"

https://viper-drones.com/industries/infrastructure-drone-use/manufacturing/#:~:text=The%20integration%20of%20drones%20into,on%20manufacturing%20is%20no%20exception.

[7] "Applications of payload drones in the telecommunications industry,"

https://gaotek.com/applications-of-payload-drones-in-telecommunications-industry/

[8] "Top 10 commercial uses for drones," April 2023,

https://www.inspiredflight.com/news/top-10-commercial-uses-for-drones.php

[9] S. D. Sree, “Drone technology in the telecom sector,” January 2023,

https://dronefluence.com/drone-technology-in-the-telecom-sector/

[10] C. Guarnera, “Connecting the world: How drones are revolutionizing the telecommunications industry,” May 2023,

https://www.bluefalconaerial.com/drones-in-telecommunications-industry/

[11] “The bright & growing future of drones in the telecom industry,”

https://pointivo.com/future-of-drones-in-the-telecom-industry/

[12] C. Guarnera, “Connecting the world: How drones are revolutionizing the telecommunications industry,” May 2023,

https://www.bluefalconaerial.com/drones-in-telecommunications-industry/

[13] “Advantages of using drones in the telecommunications industry,”

https://www.av8prep.com/aviation-library/part-107-drone/advantages-of-using-drones-in-the-telecommunications-industry

[14] “3 Reasons you should use drones for cell tower inspections,”

https://consortiq.com/uas-resources/3-reasons-you-should-use-drones-for-cell-tower-inspections

[15] “The pros and cons of drone technology,”

https://pilotinstitute.com/drone-pros-and-cons/

CHAPTER 4
BIG DATA IN TELECOMMUNICATIONS

"Consumer data will be the biggest differentiator in the next two to three years. Whoever unlocks the reams of data and uses it strategically will win."

—Angela Ahrendts

4.1 INTRODUCTION

Every day, massive amounts of data are collected using various methods, and that amount is sometimes so large that it is hard to process and analyze using traditional means. Data consumption by users and data generation by corresponding software systems are steadily growing year to year. This huge trove of data is now presented under the umbrella term "big data." We are in the big data age, where businesses in every industry must deal with vast volumes of data. Several experts and practitioners have lately emphasized the need of understanding how, why, and when big data applications may be a valuable resource for businesses seeking a competitive edge. The cloud word for big data is shown in Figure 4.1 [1].

Figure 4.1 The cloud word for big data [1].

In today's digital age, the telecom industry has become a crucial pillar for global connectivity and communication, noted for its complex infrastructure, as shown in Figure 4.2 [2]. Big data in the telecom industry encompasses a wide range of information, including customer profiles, call records, network logs, location data, social media interactions, and more. The telecom industry is a leader in big data strategy because of the vast amount of data it gathers during normal business operations. The integration of big data analytics has revolutionized various facets of the telecom industry, from enhancing customer experiences to ensuring regulatory compliance and optimizing network operations [3].

Figure 4.2 The telecom industry is noted for its complex infrastructure [2].

The world has seen a digital revolution where more and more work is being conducted online. Storing this activity has led to the concept of big data, i.e., large datasets that are otherwise difficult to manage. Because of these characteristics, big data requires new technologies and techniques to capture, store, and analyze. Telecom companies generate huge volumes of data from mobile phone usage, call detail records, server logs, network equipment, social networks, and billing. The number of data sources is growing as well. Typical sources of big data are shown in Figure 4.3 [4].

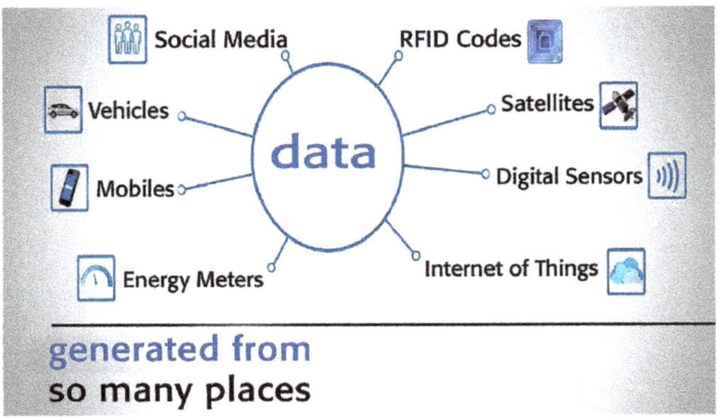

Figure 4.3 Typical sources of big data [4].

Big data and the technologies it relies on, such as AI and ML, are the driving force behind progress in the telecommunications industry. Telecommunications companies have always managed enormous volumes of network activity data and find ways they can put it to good use. Big data for telecom is the fuel that can and will drive the entire industry toward higher revenues and better customer service. Telecom companies must be able to collect this massive amount of data from different sources, analyze it, and distribute the insights to disparate databases [5].

This chapter examines the roles of big data and big data analytics in telecommunications. It begins with describing what big data is all about. It explains the characteristics of big data. It discusses the role of big data in telecommunications. It highlights the benefits and challenges of big data in telecommunications. The last section concludes with comments.

4.2 WHAT IS BIG DATA?

Big data applies to data sets of extreme size (e.g. exabytes, zettabytes) which are beyond the capability of the commonly used software tools. It involves situation where very large data sets are big in volume, velocity, veracity, and variability [6]. The data is too big, too fast, or does not fit the regular database architecture. It may require different strategies and tools for profiling, measurement, assessment, and processing.

Big Data is essentially classified into three types [7]:

- *Structured Data:* This is highly organized and is the easiest to work with. Any data that can be stored, accessed, and processed in the form of fixed format is known as a structured data. It may be stored in tabular format. Due to their nature, it is easy for programs to sort through and collect data. Structured data has quantitative data such as age, contact, address, billing, expenses, credit card numbers, etc. Data that is stored in a relational database management system is an example of structured data.

- *Unstructured Data:* This refers to unorganized data such as video files, log files, audio files, and image files. Any data with unknown form or the structure is classified as unstructured data. Almost everything generated by a computer is unstructured data. It takes a lot of time and effort required to make unstructured data readable. Examples of unstructured data include Metadata, Twitter tweets, and other social media posts.

- *Semi-structured Data:* This falls somewhere between structured data and unstructured data, i.e., both forms of data are present. Semi-structured data can be inherited such as location, time, email address, or device ID stamp.

The different types of big data are depicted in Figure 4.4 [8].

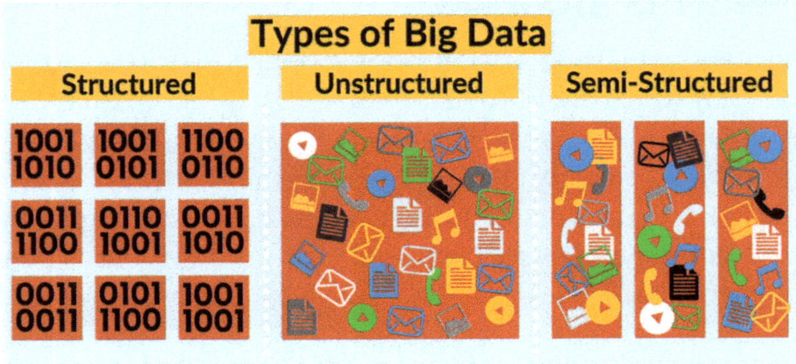

Figure 4.4 Types of big data [8].

The process of examining big data is often referred to big data analytics. It is an emerging field since massive computing capabilities have been made available by e-infrastructures [9]. Big data analytics is the application of advanced analytic techniques to large, heterogeneous data sets that comprise structured, semi-structured, and unstructured data from many sources with sizes ranging from terabytes to zettabytes.

Analytics include statistical models and other methods that are aimed at creating empirical predictions. Data-driven organizations use analytics to guide decisions at all levels. Several techniques have been proposed for analyzing big data. These include the HACE theorem, cloud computing, Hadoop, and MapReduce [10].

4.3 CHARACTERISTICS OF BIG DATA

Big data is growing rapidly and expanding in all science and engineering, including physical, biological, and medical services. Different companies use different means to maintain their big data. As shown in Figure 4.5 [11], big data is characterized by 42 Vs. The first five Vs are volume, velocity, variety, veracity, and value [2].

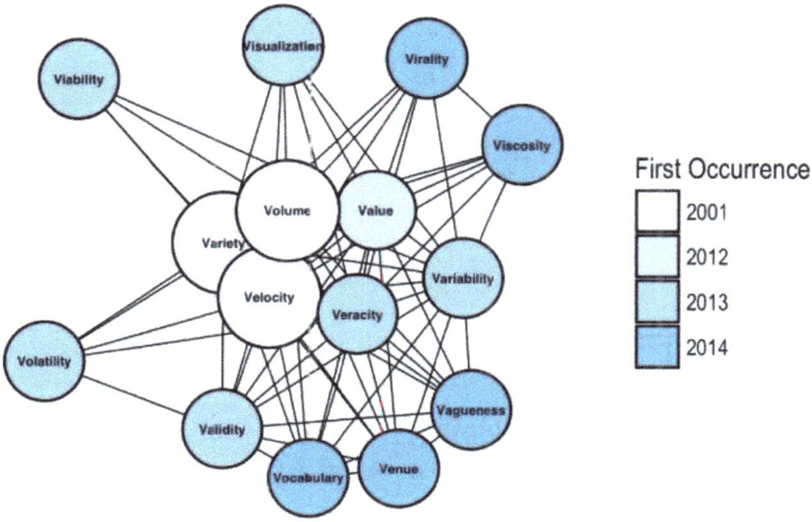

Figure 4.5 The 42 V's of big data [11].

- *Volume:* This refers to the size of the data being generated both inside and outside organizations and is increasing annually. Some regard big data as data over one petabyte in volume.

- *Velocity:* This depicts the unprecedented speed at which data are generated by Internet users, mobile users, social media, etc. Data are generated and processed in a fast way to extract useful, relevant information. Big data could be analyzed in real time, and it has movement and velocity.

- *Variety:* This refers to the data types since big data may originate from heterogeneous sources and is in different formats (e.g., videos, images, audio, text, logs). BD comprises of structured, semi-structured or unstructured data.

- *Veracity:* By this, we mean the truthfulness of data, i.e. weather the data comes from a reputable, trustworthy, authentic, and accountable source. It suggests the inconsistency in the quality of different sources of big data. The data may not be 100% correct.

- *Value:* This is the most important aspect of the big data. It is the desired outcome of big data processing. It refers to the process of discovering hidden values from large datasets. It denotes the value derived from the analysis of the existing data. If one cannot extract some business value from the data, there is no use managing and storing it.

On this basis, small data can be regarded as having low volume, low velocity, low variety, low veracity, and low value. Additional five Vs has been added [12]:

- *Validity:* This refers to the accuracy and correctness of data. It also indicates how up to date it is.
- *Viability:* This identifies the relevancy of data for each use case. Relevancy of data is required to maintain the desired and accurate outcome through analytical and predictive measures.
- *Volatility:* Since data are generated and change at a rapid rate, volatility determines how quickly data change.
- *Vulnerability:* The vulnerability of data is essential because privacy and security are of utmost importance for personal data.
- *Visualization:* Data needs to be presented unambiguously and attractively to the user. Proper visualization of large and complex clinical reports helps in finding valuable insights.

Instead of the 10V's above, some suggest the following 5V's: Venue, Variability, Vocabulary, Vagueness, and Validity) [13].

Industries that benefit from big data include the healthcare, financial, airline, travel, restaurants, automobile, sports, agriculture, and hospitality industries. Big data technologies are playing an essential role in farming: machines are equipped with sensors that measure data in their environment. Structured and unstructured data are generated in various types [14-16].

4.4 BIG DATA IN TELECOMMUNICATIONS INDUSTRY

Big data analytics is revolutionizing the telecommunication industry by enabling targeted marketing and enhancing customer behavior insights. This technology continues to be integral in shaping the future of telecom, providing innovative solutions. The telecommunications industry is making significant strides in technological advancements, and big data analytics (BDA) is playing a crucial role. Figure 4.6 shows the big data impacts on telecommunications industry [17], while Figure 4.7 illustrates big data analytics use cases in telecommunication industry [18].

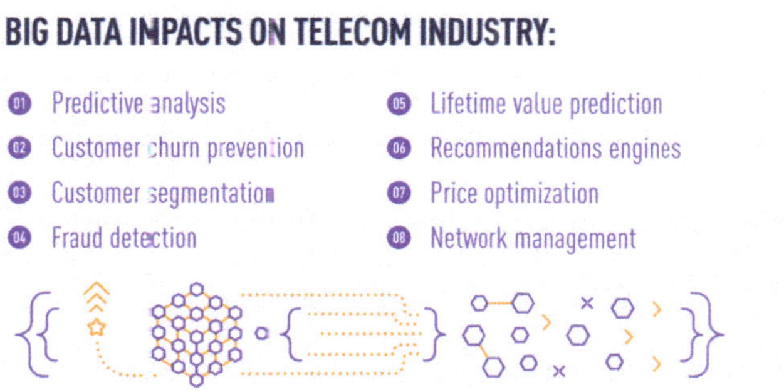

Figure 4.6 Big data impacts on telecommunications industry [17].

Figure 4.7 BDA use cases in telecommunication industry [18].

Telecom startups and companies can take advantage of big data analytics in the following main areas [19,20]:

- *Customer Experience:* Big data analytics in the telecom industry can help companies improve customer experience. Maintaining a current customer relationship is considerably cheaper than acquiring a new one, and this improves the company's profitability in the long term. Processing big data can help with personalizing the customer experience. It also empowers telecoms to streamline their service portfolios, design, and implement new features, and provide the best customer support possible. For example, a telecom company can collect information from users regarding various service issues and then create an automated chatbot that will assist customers to resolve these issues immediately whenever possible. Figure 4.8 shows a typical Telco customer [21].

Figure 4.8 A typical Telco customer [21].

- *Network Optimization:* Network optimization stands as a cornerstone for telecom companies aiming to enhance service quality, maximize efficiency, and meet the ever-growing demands of users. It entails complex and quick evaluation of large volumes of data in real-time mode. Big data helps telecom operators in optimizing the quality of provided services by integrating network optimization. When combined, network telemetry, CDRs, data statistics, and equipment alerts enable communications service providers (CSPs) to set up and maintain effective network self-diagnostics and self-configuration tools. With big data analytics, telecom companies can monitor their network performance. For example, companies can record and gather issues raised by users from a particular area and look for solutions to resolve these issues, such as improving the connectivity.

- *Operational Analysis:* Companies can use big data for both analyzing and modifying operations as the need arises. Analysis can also be used to monitor the telecom's usage of resources and thereby prevent waste, which leads to savings of both resources and money. Real-time operational analysis also helps set the timetable for data updates and define other parameters so a company can best adapt the data analysis system to their unique business requirements.

- *Data Monetization:* Big data collected by telcos can be sold to or shared with third parties interested in monetizing it. Big data allows companies to gather a lot of user-related information, like demographics and location info, network and application usage, details about used devices, various preferences, and more. After processing, this data has value that extends beyond the companies that gathered them, opening a possibility for data monetization. Telecom companies can sell this data – without violating users' privacy – to companies from other industries like healthcare, financial services or advertising. For example, insurance companies, marketing agencies, banks, and other financial institutions may be interested in the behavior of a particular cohort of users, which will help them optimize their service offerings, and, thus, boost big data monetization in telecoms.

- *Fraud Detection:* Most fraud cases in the telecom industry are illegal access, fake profiles, authorization, cloning, or behavioral fraud. Each of these frauds affects the relationship established between the company and the customer directly. Therefore, fraud detection tools and techniques are needed. Big data analytics helps to monitor behaviors in real-time, thus preventing fraud. This technique has extremely high efficiency because it allows an almost real-time response to any suspicious activity. AI tools working on top of layers of accumulated and constantly updated relevant big data help telcos live up to high security standards and retain their customer base. Cybercriminals are raking millions of dollars by employing sophisticated scams across different geographies and vanishing even before the communication service provider knows. The revenue lost through Telco fraud amounts to billions of dollars each year.

- *Price Optimization:* The number of customers in the telecom industry customers is growing extremely fast. Therefore, pricing optimization becomes an important factor. The telecom industry is rife with cutthroat competition among different service providers, competing to have the largest subscriber market share. Product pricing is a key factor for operators competing to gain more subscribers. Telecoms use big data technologies to review several real-time metrics and settle on a particular product's price. For example, an operator may announce that users would get free minutes daily for calls made at a particular time frame under a particular voice bundle subscription. This sort of product offering comes about after analyzing data sources and then deriving actionable insights from them. The providers can set the optimal service prices to gain and retain customers.

- *Customer Churn Prevention:* Churn rate, also known as attrition rate, refers to the percentage of customers that stop using a particular service within a forecast period. In a recent study, the telecom industry has a churn rate of 21 percent, closely linked to poor customer service. Engaging new customers takes a lot of time and effort, and so does retaining customers. Companies have to analyze customers' behaviors and take corresponding actions to prevent customer churn. Telecom data analytics use cases give insights into this data, help disclose customers' feelings regarding services they receive, immediately address satisfaction issues, and prevent churn.

- *Product Development:* Developing a product is a complex process requiring proper control and management. It is a time-consuming. However, the integration of big data analytics ensures high-performing and quality products for the consumers, as per their needs and requirements. As a result of telecom data analysis, companies can develop data-driven products with internal feedback and marketing strategies. Big data analytics use cases in telecom companies can greatly assist with data-driven product development, internal feedback, and marketing intelligence.

4.5 BENEFITS

Big data can help the telecom industry in a myriad of ways. Big data is helpful for telecom operators in optimizing the quality of provided services by integrating network optimization. Companies can harness big data's potential by investing in custom data solutions that will help them with optimizing analytics to increase profitability. Big data opens up a sea of opportunities in the telecom industry, including improved customer experience, fraud detection, network optimization, risk mitigation, data monetization, and more. If done right, big data is the fast route to customer retention, growing subscriber base, and profitability. Other benefits include [22]:

- *Improved Customer Experience:* Competition within the telecom industry is fierce. Rival companies are always seeking opportunities to win over clients in ways besides decreasing prices or offering bonuses. And that is what makes big data so appealing; it opens new possibilities in improved customer experience, more personalized marketing, reducing churn rate, and generating new streams of revenue. Big data allows telecom companies to gather and categorize vast amounts of user information, enabling them to personalize customer experiences. By analyzing customer data, telecom companies can provide proactive assistance, resolve issues quickly, and offer tailored services.

- *Targeted Products:* Out of the customer micro-segments, you can tailor targeted and relevant product offerings that entice the customer. By leveraging customer behavior patterns, billing information, and issue resolution data, telecom companies can not only resolve customer problems and improve services but also target customers with relevant offerings. For example, a customer who makes frequent calls to a specific country can be targeted with a specific marketing campaign in real-time.

- *Predictive Capacity:* Big data can also help with developing predictive capacity models to predict network problems that may arise in the future. This includes predicting peak network loads and allocating adequate network resources to relieve congestion.

- *Personalized Offers:* Telecoms can use big data to segment customers into special categories based on their network usage data. This helps them present deeply personalized offers and services to the respective customer micro-segments, promoting customer retention and halting growing churn trends.

- *Targeted Marketing:* Under target marketing, the companies utilize big data network analytics in telecom to provide consumers with customized services on the basis of their purchase history, preferences, and feedback. Technology also helps companies identify customer behavior by evaluating their service usage.

- *Preventive Diagnostics:* Preventive diagnostics involve identifying the pattern of the service behavior to avoid system failures. Business analytics in telecom industry performs preventive diagnostics to analyze consumer intention

- *Enhanced Customer Experience:* The ultimate goal of telecommunication is to build a positive customer experience. By using data analytics systems, companies can use big data analytics system that uses comprehensive analysis to provide creative workflows using infographics. Besides, it also allows the industries to personalize the customer experience by sorting and separating the data into diverse categories.

- *Predictive Analysis:* The predictive analysis applied by telecom companies uses historical data to predict future behaviors and get valuable insights into customer data. With gathered insights, companies can become faster, more efficient, and better. It also helps with making data-driven decisions. For example, by analyzing customer preferences, companies can have a better understanding of each customer.

- *Customer Segmentation:* The content and strategy need to be tailored to the specific market. Therefore, telecom companies segment their customers and then target campaigns accordingly. Segmentation and targeting help with predicting customers' needs, preferences, and reactions to services and products.

- *Predicting Churn:* Churn refers to is a measure of the number of individuals or items moving out of a collective industry over a specific period making it a real issue of any business or industry. This can be due to several reasons like quality of service, network issues, social media trends, availability of other better options, sudden price hikes, unresolved queries. Understanding such scenarios with the data available can always be a preventive way of reducing churn.

- *Network Security:* Network security is of paramount importance in the telecom industry, given the sensitivity and value of the data transmitted over telecommunications networks. Big data analytics serves as a powerful tool for identifying and mitigating security threats, enhancing cybersecurity measures, and safeguarding network integrity and data confidentiality

- *Data Governance:* Data governance is very essential to protect the data and assets of telecom operators. It refers to the policies and procedures adopted in order to define how data assets could be accessed and manipulated and by whom in the organization. A common mistake that telecom operators makes is that they confuse data quality management, which comprises activities for the improvement of data quality, and data governance. This can be explained by the close connection between data governance and data quality management results from a data perspective as a company asset.

4.6 CHALLENGES

Like any other industry, the telecom industry is faced with many challenges. Competition in the telecommunications market has never been more vicious. The industry has every chance of beating these challenges. These challenges include the following:

- *Predicting Churn:* Churn, the number of customers leaving a service, poses a significant challenge for telecom companies. Big data analysis allows companies to understand the reasons behind churn, such as service quality, network issues, social media trends, price changes, and unresolved queries.

- *Detection of Fraud:* The most common types of fraud in the telecom business include unauthorized access, false profiles, authorization, duplication, behavioral fraud, and more. Fraud has a direct impact on the relationship that has been built between the organization and the user. As a result, big data analytics aids in real-time monitoring and fraud prevention. This technology is extremely efficient since it enables real-time response to any suspicious activity.

- *Talent Capability:* Talent capability in the context of analytics refers to the expertise of data analyst or scientists in executing tasks related to big data analytics. This capability is one of the defining components in creating and sustaining a competitive advantage. A close relationship between data scientists and other people within the organization can be very important. Figure 4.9 shows big data analytics talent capability pillar [23].

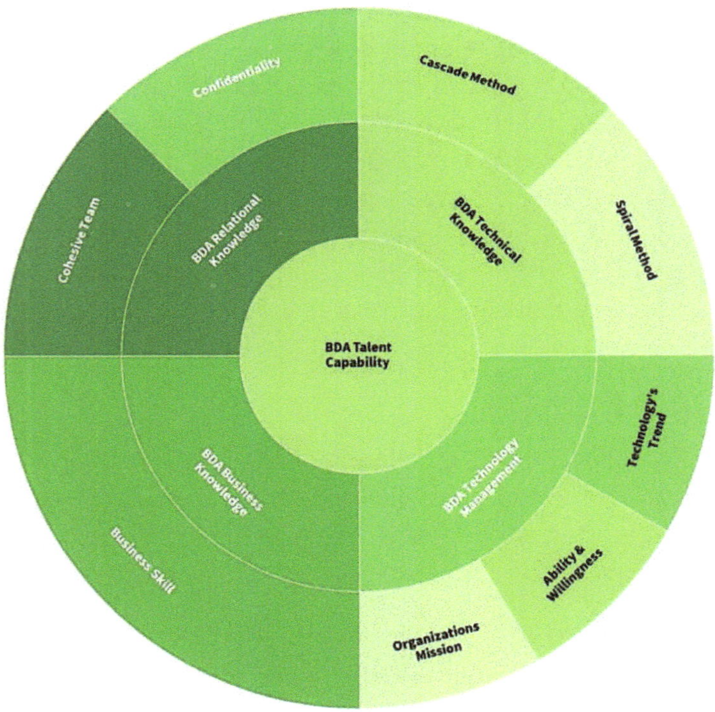

Figure 4.9 Big data analytics talent capability pillar [23].

- *Skills Shortage:* Lack of talented resources is a key challenge. The most challenging in BDA (big data analytics) projects is finding a qualified team. This can be explained by the fact that BDA is always considered as new technology compared to business intelligence, which most organizations have built over decades. Advanced analytics requires staff with deep knowledge in different domains, from data science to worldwide privacy laws, along with an understanding of the telecommunications business.

4.7 CONCLUSION

To cope with all the changes, companies have to stay alert to the newest trends in big data analytics in the telecom industry. Today, the application of big data in the telecom industry is multi-fold. Telcos can leverage big data technologies to turn vast amounts of raw data into actionable insights.

Big data analytics in the telecom industry refers to the process of analyzing large volumes of data generated by telecommunications networks and services to extract valuable insights. It is revolutionizing the telecommunication industry by enabling targeted marketing and enhancing customer behavior insights. The application of big data in telecom companies is crucial for success in the current and future telecom industry. The synergy between big data analytics and emerging technologies like artificial intelligence and machine learning continues to redefine the possibilities within the telecom sector, paving the way for innovation and growth. More information about big data in the telecommunications industry is available from the books in [24-26] and this related journal: Journal of Big Data.

REFERENCE

[1] R. Delgado, "The challenges of bringing BYOD to the military,"

https://socpub.com/articles/the-challenges-of-bringing-byod-to-the-military-11272

[2] M. Atherton, "Telecom infrastructure vulnerabilities: Assessing and enhancing resilience," January 2024,

https://www.linkedin.com/pulse/telecom-infrastructure-vulnerabilities-assessing-dr-marlon-atherton-c2rpe/

[3] A. Pareek, "Global examples of telecom companies using big data analytics," July 2022,

https://customerthink.com/global-examples-of-telecom-companies-using-big-data-analytics/

[4] J. Moorthy et al., "Big data: Prospects and challenges," *The Journal for Decision Makers*, vol. 40, no. 1, 2015, pp. 74–96.

[5] M. N. O. Sadiku, P. A. Adekunte, and J. O. Sadiku, "Big data in telecommunications," *International Journal of Trend in Scientific Research and Development*, vol. 8, no. 6, November-December 2024, pp. 243-252.

[6] M. N.O. Sadiku, M. Tembely, and S.M. Musa, "Big data: An introduction for engineers," *Journal of Scientific and Engineering Research*, vol. 3, no. 2, 2016, pp. 106-108.

[7] "The complete overview of big data,"

https://intellipaat.com/blog/tutorial/hadoop-tutorial/big-data-overview/

[8] R. Allen, "Types of big data | Understanding & Interacting with key types (2024),"

https://investguiding-com.custommapposter.com/article/types-of-big-data-understanding-amp-interacting-with-key-types

[9] P. Baumann et al., "Big data analytics for earth sciences: the

earthserver approach," International Journal of Digital Earth, vol. 19, no. 1, 2016, pp.3-29.

[10] X. Wu et al., "Knowledge engineering with big data," *IEEE Intelligent Systems*, September/October 2015, pp.46-55.

[11] "The 42 V's of big data and data science,"

https://www.kdnuggets.com/2017/04/42-vs-big-data-data-science.html

[12] P. K. D. Pramanik, S. Pal, and M. Mukhopadhyay, "Healthcare big data: A comprehensive overview," in N. Bouchemal (ed.), *Intelligent Systems for Healthcare Management and Delivery*. IGI Global, chapter 4, 2019, pp. 72-100.

[13] J. Moorthy et al., "Big data: Prospects and challenges," *The Journal for Decision Makers*, vol. 40, no. 1, 2015, pp. 74–96.

https://www.grandviewresearch.com/industry-analysis/industrial-wireless-sensor-networks-iwsn-market

[14] A. K. Tiwari, H. Chaudhary, and S. Yadav, "A review on big data and its security," *Proceedings of IEEE Sponsored 2nd International Conference on Innovations in Information Embedded and Communication Systems*, 2015.

[15] M. B. Hoy, "Big data: An introduction for librarians," *Medical Reference Services Quarterly*, vol. 33, no 3. 2014, pp. 320-326.

[16] M. Viceconti, P. Hunter, and R. Hose, "Big data, big knowledge: Big data for personalized healthcare," *IEEE Journal of Medical and Health Informatics*, vol. 19, no. 4, July 2015, pp. 1209-1215.

[17] K. Goworek, "The big impact of big data on the telecom industry," January 2021,

https://tasil.com/insights/big-data-in-telecoms/

[18] S. Srivastava, "How telecom companies use big data analytics – Top 10 use cases," August 2024,

https://appinventiv.com/blog/big-data-analytics-for-telecom-industry/

[19] “Big data in the telecom industry: A golden harvest on the digital fields,” June 2024,

https://startups.epam.com/blog/big-data-analytics-for-telecommunications

[20] R. Makarchuk, “Big data in the telecom sector: Trends, use cases & case studies,”

https://intellias.com/big-data-telecom/

[21] “Big data analytics in telecom industry. Use cases,” April 2024,

https://addepto.com/blog/the-role-of-big-data-in-the-telecom-industry/

[22] A. Kumar, “10 Use cases of data analytics in telecom industry,” September 2024,

https://www.appventurez.com/blog/big-data-analytics-in-telecom-industry

[23] H. Keshavarz et al., “The value of big data analytics pillars in telecommunication industry,” *Sustainability*, vol. 13, no. 13, June 2021.

[24] Y. Ouyang and M. Hu (eds.), *Big Data Applications in the Telecommunications Industry*. Information Science Reference, 2016.

[25] R. He and Z. Ding (eds.), *Applications of Machine Learning in Wireless Communications (Telecommunications)*. The Institution of Engineering and Technology, 2019.

[26] A. K. Bashir et al. (eds.), *Implementing Data Analytics and Architectures for Next Generation Wireless Communications*. IGI Global, 2021.

CHAPTER 5
INTERNET OF THINGS IN TELECOMMUNICATIONS

"And just like any company that blissfully ignored the Internet at the turn of the century, the ones that dismiss the Internet of Things risk getting left behind."

—Jared Newman

5.1 INTRODUCTION

Technology is changing fast, and the Internet of things (IoT) is a big part of it. IoT is a network of everyday physical objects or devices. These objects can be anything, from your smartphone, smartwatch, or smart home devices like thermostats and lights, to larger things like smart cars and industrial machines. IoT is used in smart homes, offices, self-driving cars, and even security systems. IoT has become a commonplace practice and are being permanently improved and expanded. IoT solutions are actively implemented in various sectors around the world, especially in the telecom industry in recent years.

Retail, agriculture, healthcare, manufacturing, and utilities are just a few of the industries that are using IoT solutions. The number of devices connected to the Internet of things worldwide is forecast to triple by 2030.

The telecom infrastructure stands at the forefront of technological evolution, and is noted for its complexity, as typically shown in Figure 5.1 [1]. Everything from work to play is shaped by modern telecom's unified voice, video, and information architecture.

Figure 5.1 Complex telecom infrastructure [1].

Telecommunications play a key role in creating an interconnected world through the Internet of things. They provide the essential infrastructure and services needed to connect IoT devices to each other, to the Internet, and to other systems. The effects of IoT adoption will mean that telecoms have an increasing role to play in areas like healthcare, manufacturing, city infrastructure, factories, and integrated home environments. Industry giants like Vodafone, AT&T or Verizon, are playing a crucial role in shaping the future of IoT and staying ahead of the game. They already offer narrowband IoT networks that can ensure low-cost, stable signal coverage for connected devices.

The Internet of things is a concept that refers to a network of everyday physical objects or devices that are connected to the Internet and can communicate with each other and with end-users. Today, IoT technology has entered practically all industries, and it is hard not to note its impact on the telecommunications sector. The implementation of IoT-based initiatives is a must for telecom businesses striving to embrace the digital age, getting a strong competitive advantage, and making the most of the possibilities that modern technology offers. The Internet of things strategies for telecommunications have been completely changing the technological landscape. It is revolutionizing the telecommunications industry, having a huge impact on all business processes [2].

This chapter examines the roles of IoT in the telecommunications industry. It begins with explaining the Internet of things. It discusses the role of IoT in telecommunications. It highlights the benefits and challenges of IoT in telecommunications. The last section concludes with comments.

5.2 OVERVIEW ON INTERNET OF THINGS

The Internet began with some military computers in the Pentagon called Arpanet in 1969. It expanded throughout the 1980s as a set of four parallel military networks, each at a different security level. The core technology which gives the Internet its particular characteristics is called Transmission Control Protocol/ Internet Protocol (TCP/IP), which is essentially a set of rules for communication [3].

Internet of Things (IoT) is a worldwide network that connects devices to the Internet and to each other using wireless technology. IoT is expanding rapidly and it has been estimated that 50 billion devices will be connected to the Internet by 2020. These include smart phones, tablets, desktop computers, autonomous vehicles, refrigerators, toasters, thermostats, cameras, alarm systems, home appliances, insulin pumps, industrial machines, intelligent wheelchairs, wireless sensors, mobile robots, etc. The basic idea of IoT can be conceived as a representation of various A's is shown in Figure 5.2 [4]. Figure 5.3 illustrates typical applications of the Internet of things [4].

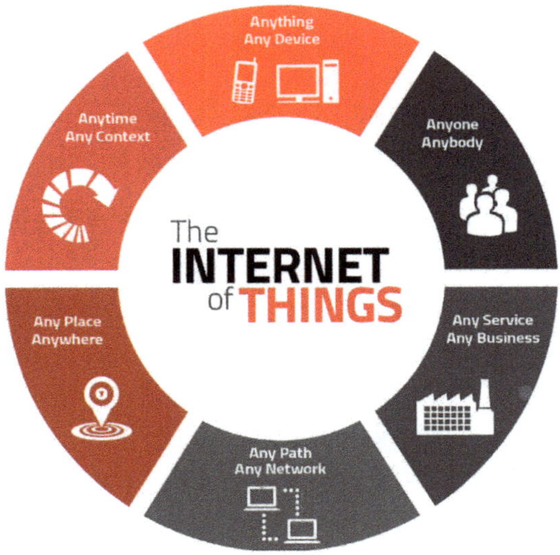

Figure 5.2 IoT can be conceived as a representation of various A's [4].

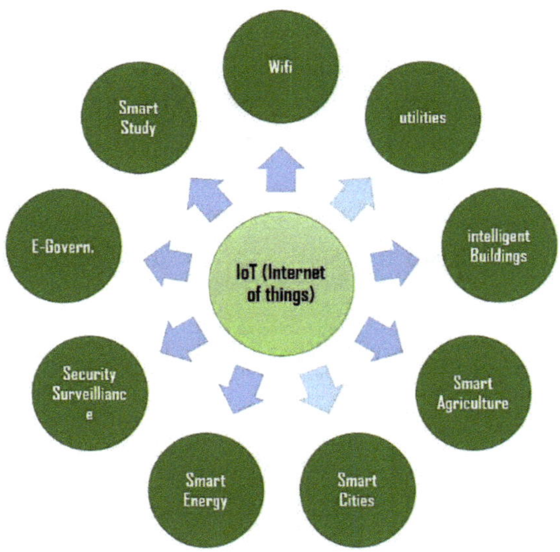

Figure 5.3 The Internet of things [5].

There are four main technologies that enable IoT [5]: (1) Radio-frequency identification (RFID) and near-field communication, (2) Optical tags and quick response codes: This is used for low cost tagging, (3) Bluetooth low energy (BLE), (4) Wireless sensor network: They are usually connected as wireless sensor networks to monitor physical properties in specific environments. Communications technologies in Internet of things are portrayed in Figure 5.4 [6].

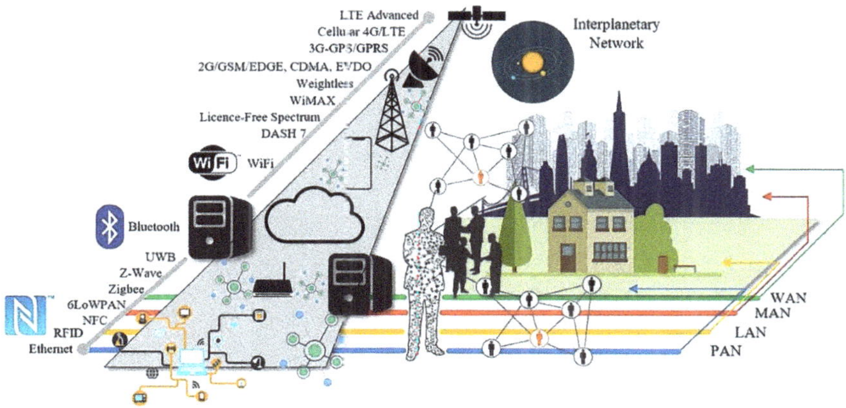

Figure 5.4 Communications technologies in Internet of things [6].

IoT technology enables people and objects to interact with each other. It is employed in many areas such as smart transportation, smart cities, smart energy, emergency services, healthcare, data security, industrial control, logistics, retails, structural health, traffic congestion, manufacturing, and waste management. The Internet of things is extensively developed world-wide with a focus on civilian applications such as electric power distribution, intelligent transportation, healthcare, industrial control, precision agriculture, environmental monitoring, etc.

5.3 INTERNET OF THINGS IN TELECOMMUNICATIONS INDUSTRY

The key role of IoT is that it improves everything around us, making objects smart, connected, and able to interact with each other. Internet of things technologies transform the role of mobile network providers in ensuring the connection between users and devices. Telecom companies use the Internet of things (IoT) to deliver a collection of products and services that bring additional value to their existing networks. Figure 5.5 shows major IoT use cases in telecom [7]. Telecom companies can take advantage of IoT in the following key ways [8,9]:

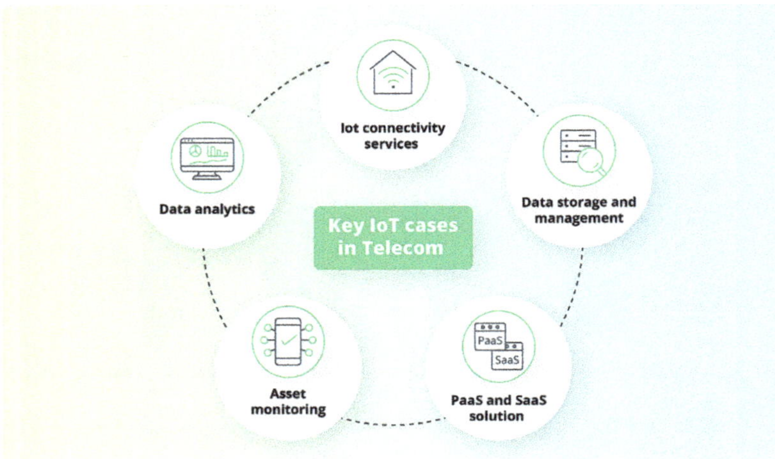

Figure 5.5 Major IoT use cases in telecom [7].

• *IoT Connectivity:* A first and foremost IoT implementation option for telecom providers is to offer connectivity services to those users who have IoT devices. It is offering connectivity services to users who already own IoT devices. The IoT is bringing a new era of interconnectivity to our world. The basic principle behind IoT is connectivity between web-enabled smart devices and the cloud and between the devices themselves. An IoT connectivity platform is responsible for managing devices and improving the quality of connecting the customer's infrastructure to the telecom network. The Internet of things in telecom provides seamless and secure communications, and the challenge is for industry leaders to tune their physical assets to offer IoT-enabled connectivity platforms. Providing the connectivity required for IoT networks will define future telecom operations.

• *Data Analytics:* IoT initiatives provide telecommunications businesses with an enormous amount of data, enabling telcos to create information pools or apply this data for analytics. Today, telcos have unique opportunities to use analytics combined with data from IoT devices. By implementing Internet of things initiatives, the telecommunication business can obtain huge amounts of vital data. Analytics, combined with data generated by IoT devices and supplemented with algorithms based on artificial intelligence and machine learning, also facilitates tracking user behavior, allowing telecom providers to carry out diagnostics and develop forecasting models.

- *Data Storage and Management:* Since IoT creates lots of data, telecommunications companies help manage it. They store, process, and analyze the data so that businesses can make smart decisions based on the information. Telecom companies can also store and process data gathered from clients' IoT devices. A customer runs a business application to access any needed information on the centralized platform. In this scenario, the provider is responsible for all backend processes including data storage, filtering, classification, and cleansing.

- *Asset Monitoring:* Monitoring assets remotely is another popular use case of IoT in the telecom sector. By using an IoT platform, telecom companies can connect their diverse physical assets to the cloud and remotely manage their operations, investigate malfunctions, run firmware upgrades, and keep track of inventory. For many industries, the location of physical assets such as vehicles, orders, or even patients is critical to track in real time via mobile apps. Using extended communications networks and cloud computing, telcos can monitor asset movements by collecting signals from embedded sensors and transmitting them to industry-tailored applications.

- *PaaS and SaaS Solutions:* Platform as a service (PaaS) and software as a service (SaaS) solutions are among the most profitable ways for telecom companies to generate revenue. The PaaS model enables clients to independently develop their own solutions by accessing cloud-based IoT platforms. SaaS applications offered by telecom operators can be tailored for practically any industry and bring exceptional benefits.

- *Location Services:* Mobile networks and IoT sensors help realize advanced and improved location capabilities, which is a necessary feature of smart cities. These IoT sensors can be integrated into public transport infrastructure to provide users with optimized routes, find the nearest stops, and pick transport in real-time.

- *Autonomous Vehicles:* Autonomous driving is being actively explored by numerous carmakers all over the world. Connected vehicles, autonomous driving, and the automotive infrastructure as a whole are some of the key areas for the Internet of things and 5G mobile technologies. For cars to be completely autonomous, their environment must be equipped with sensors to send signals that can be interpreted in real-time to make driving secure and reliable. The introduction of IoT devices with LTE-V solutions allows cars to interact, notify each other about a lane change, detour, or emergency braking, prevent accidents, and communicate with the surrounding road environment.

- *Equipment Monitoring:* Telecom capabilities provide support for communication between people around the world and let businesses access global markets. Telcos use a huge amount of equipment to ensure high-quality and consistent service, and its failure is one of the most common reasons for telecommunication breakdowns. The equipment must always stay in working condition and operate flawlessly around the clock, and any failures must be corrected instantly. IoT technologies help meet this challenge. Their sensors are able to collect and analyze data from cell towers, and also allow identifying the problems, responding to them, and carrying out surveillance in real-time.

- *Hazard Detection:* Equipment failure is not the only reason that prevents the uninterrupted provision of telecom services. Among other critical threats are global catastrophes that telcos must also immediately respond to since the lack of control over such hazards can cause great damage to equipment. The capabilities of the Internet of things allow communication providers to effectively manage facilities, monitor incidents, and detect disasters in time so that operators can perform an emergency shutdown of equipment to prevent its total breakdown.

- *Security:* One more critical point in terms of security at remote sites is the exposure of expensive and critical equipment to the possibility of theft. This means that telcos must implement extremely reliable and robust protection systems. IoT-based smart cameras become the best solution in terms of physical security.

- *Healthcare Industry:* The healthcare industry is one of the leading and most promising for telecommunications. This industry is one that is actively adopting IoT solutions and devices. The telecom companies are building ties with healthcare institutions for ensuring high-quality, fast, and cost-efficient connections. The Internet of things not only facilitates the process of monitoring the condition of patients but also allows the introduction of extra remote healthcare services. Other applicable options include remote medical diagnostics, automated incident control systems, and streamlined healthcare delivery. The implementation of IoT solutions in the medical industry is costly due to the increased privacy and complexity of the systems, but it is definitely something that will pay off in the near future. The most significant applications of IoT within remote healthcare are in remote asset management platforms, remote healthcare diagnostics devices and solutions, and smart incident management systems. Due to data sensitivity and the sophistication of technology, deploying IoT technologies in the medical sector requires substantial investment.

- *Smart Cities:* The main objective of smart cities is to optimize how a city functions and promote economic growth while also improving the quality of life for citizens by using smart technologies. This could include smart surveillance cameras, waste management, and predictive and proactive healthcare. It is difficult to overestimate the role of telecommunications in smart home and smart city technologies. The core of their implementation consists of 5G and the narrowband IoT. The idea of smart cities has developed around the world, with technology in use from New York to Copenhagen. Cities are using data from smart devices to assist with power consumption and distribution, traffic management, and environmental efforts. If in smart home technology we are talking about managing security, lighting, electronics, and others, then in the development of smart cities, the role of telco and IoT comes down to creating an efficient, inclusive, and interconnected ecosystem. In all these cities, telecom partners have helped implement the necessary 5G and narrowband IoT networks. The success of these initiatives means that more cities will be looking to modernize with IoT integration. Figure 5.6 shows a typical smart city [10].

Figure 5.6 A typical smart city [10].

- *Smart Homes:* Smart home systems have been around for a while, but they have been modified and enhanced a lot. Today, citizens are using sophisticated devices based on latest technology for their daily usage at homes. We know that this ecosystem is heavily dependent on high-speed Internet. Telcos are playing an important role in their development. The IoT version of a smart home relies on interconnectivity on a network level. This gives telecom providers a unique opportunity in smart home networks. Rather than relying on separate smart devices with different communication standards, the devices operate on a smart network. This is managed by the telecom provider in the cloud and the user accesses it all with an app on their smartphone. IoT and telecommunications play an important role in tasks and processes involved in smart homes, such as management of lighting, temperature, appliances, security, and others. Deutsche Telecom, AT&T, and Comcast all have their own versions of smart home network services. Figure 5.7 depicts a typical smart home [7].

Figure 5.7 A typical smart home [7].

5.4 BENEFITS

The benefits and opportunities of the Internet of things allow telecommunication enterprises and start-ups to expand their existing services and products, as well as introduce new ones that go far beyond their basic offerings. IoT technologies are responsible for automation, innovation, and transformation. The IoT affects both internal and external processes, improving customer satisfaction, enhancing the quality of service, creating new sources of income, increasing return on investment, and completely transforming traditional approaches in telecom. Other benefits include [8]:

- *Operational Efficiency:* IoT-based technology makes it possible to modify the process of collecting data in real-time to optimize performance, decrease resource requirements, and avoid errors caused by the human factor. IoT solutions can affect various critical processes. For example, they can improve customer experience and level of satisfaction.

- *Higher Speed:* Where connectivity and speed are mission critical for a remotely-operated device and the network drops, becomes slower or lags, the results would be loss of time, money and equipment. With IoT technologies, telecommunication service providers can also leverage the advantages of advanced bandwidth and increased speed of data exchange. The interaction of IoT and telecom companies leads to accelerated speed and advanced bandwidth. IoT protocols and 5G interconnection are able to broadcast information from thousands of devices to numerous users without slowing down communication speed or reducing bandwidth. The introduction of IoT innovations is always accompanied by an increase in the energy efficiency of telcos and the ability to serve millions of connected devices.

- *Customer Satisfaction:* By applying the Internet of things and using it to collect and analyze large amounts of data generated by telecom operators, businesses can identify current tendencies and determine customer behavior, which will let them implement better solutions in terms of marketing, sales, and production, helping to attract new customers and retain existing ones. Hence, the use of IoT in the telecom industry can optimize the user experience and allow telcos to dominate the industry in terms of customer satisfaction.

- *Monitoring:* Telecommunications companies need to maintain constant control over their cell towers, and it is expensive and time-consuming to monitor the state and functioning of each tower manually. Smart monitoring in real-time can predict machine errors and reduce downtime from malfunctions. Leading businesses in the telecom industry are using IoT to trace products, monitor customer satisfaction levels, keep track of employee and user engagement, and more. For example, the Internet of things in telecom can help prevent the spread of infections by following the location of an infected patient and their contacts.

- *Enhanced Security:* Safety must be given priority. Ideas like smart cities are exactly what is fueling the growing demand for Internet of things security. The implementation of IoT solutions in smart cities will definitely provide enhanced security. One of the leading tendencies in increasing security, enhancing privacy, and providing IoT data protection is blockchain technology, which is embedded into IoT networks to provide improved confidentiality and safety.

- *Enhanced Safety:* Every enterprise that has remote facilities or sites knows that the risks related to these objects are extremely high. There are a lot of weather factors and conditions (like fire) that cannot be controlled but have a serious impact. The Internet of things in telecommunications can monitor the safe operation of equipment, giving the company the ability to take timely preventive measures and avoid serious damage.

- *Service Creation:* New digital services need to be created as IoT applications develop. These services need to be broad in scope and will require a lot of forward planning. There is huge revenue potential in establishing these services, though, as IoT platforms become more popular.

- *Service Provision:* Telecom companies are ideally suited to becoming IoT platform providers. Their ability to act as service providers goes beyond networks. They have the capacity to process large amounts of data and to develop the existing digital infrastructure that is required.

- *Network Development:* The most natural role for telecom providers is network development. This involves the creation and maintenance of network infrastructure for 5G networks that can handle IoT data. This is actually not dissimilar to what many telecom providers are currently doing.

- *Roaming Capabilities:* The telecoms that support and use IoT devices have roaming capabilities where devices can still connect to a cellular network in geographic areas where the primary telecom does not have any infrastructure. AT&T, Orange, Vodafone, and Verizon all claim they have roaming capabilities for their IoT devices because of the SIM cards' capabilities in the devices.

- *Increased Revenue:* Due to the implementation of IoT services, telecom operators can introduce innovative value-added services for their clients. Growing customer satisfaction lets telecommunication companies increase their profits.

- *Business Opportunities:* Another benefit of IoT is that it creates more job opportunities in the telecom sector as a result of the deployment of new technology and the creation of new alternatives. The convergence of telecommunications and the Internet of things is opening up new business opportunities. Companies can leverage advances in telecommunications to create new products, services, and solutions that can increase efficiency, streamline processes and create new revenue streams.

• *Energy Savings:* AT&T was able to reduce the energy consumption by 9 million kWh per year thanks to energy management using the Internet of things, saving almost one million dollars. Figure 5.8 shows some of these benefits [11].

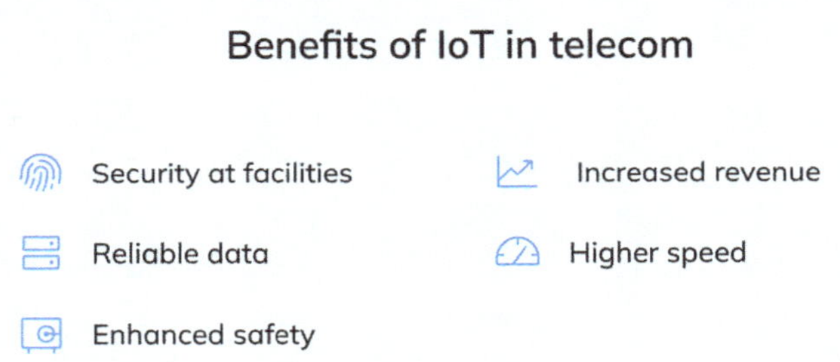

Figure 5.8 Some benefits of IoT in telecom [11].

5.5 CHALLENGES

The Internet of things represents enormous potential for telecommunications companies, but also poses a number of challenges that need to be addressed. In spite all its promise, IoT for telcos is not that easy to adopt. The undeniable advantages and endless opportunities go hand in hand with inevitable challenges telcos face when trying to implement IoT solutions. Facing the competition on their own is not a viable option for any telecom operator. Other challenges include [8]:

• *Availability:* IoT is usually utilized to facilitate information anywhere at any time. For IoT environments to become fully functional, round-the-clock availability will be vital. The real-life applications of IoT in environments like smart factories and healthcare environments cannot afford unplanned downtime. The Internet of things is typically used for the constant and uninterrupted transmission of information, capturing user needs. Thus, accessibility is a vital challenge for the Internet of things, and guarantees of high availability of all applied telecom devices are necessary. To ensure accessibility, the best way out is to use these fallback mechanisms.

- *Reliability:* One of the critical points of the Internet of things in telecom is to ensure a reliable and uninterrupted user experience. And this will happen not only by providing reliable information but also by getting used to the changing circumstances of the environment. Absolute reliability must be provided and assured.

- *Security:* Ensuring absolute security and complete confidentiality is one of the key priorities for IoT in telecommunications. With so much interconnectivity, security has become a primary concern for users and businesses. That means that reliable networks with built-in redundancy failsafe will need to be developed within smart devices to eliminate the issue of network disruption. The problem here is that the memory devices in the IoT are only capable of storing a limited amount of data, which means that some of the information is stored on other resources remotely. Such a type of storage of important data is definitely vulnerable in terms of privacy and reliability. The secrecy and security points are not solved within the technology of the Internet of things.

- *Privacy:* IoT devices store only a small amount of data locally, while the rest is saved on the cloud. Users worry about their personal data getting into the wrong hands, so telecom companies must protect their customers' privacy. Telecom companies should implement strong encryption techniques to secure the communication between IoT devices and the cloud.

- *Interoperability:* IoT technology envisages connecting various devices. This means that interoperability is a must, regardless of device type. The Internet of things in telecom must provide services equally to all connected devices.

- *Scalability:* It is important to consider the scalability of connected devices in such a way as to ensure fast processing of all data. The telecom sector is an incredibly fast-paced environment where the underlying infrastructure has to continuously scale alongside the continuous growth of the number of connected devices. As organizations add more connected devices, IoT scaling gets buried under an avalanche of new forms of data from devices such as sensors, gateways, routers, and cameras.

- *Compatibility:* Various devices should be connected and should be able to communicate with each other. Compatibility between products, cloud storage solutions, and business tools will be vital in large environments using multiple smart devices. This presents a significant challenge in a market saturated with different tech businesses pursuing their own standards. Good interoperability is a prerequisite for the implementation of the Internet of things in telecommunications. It must be maintained between various products that are permanently connected. The fact is that with the growth of the IoT, different technologies are battling to be the standard. Such competition obviously leads to integration problems. Users must update their gadgets and install all fixes to maintain constant compatibility. Diverse industries must work together to ensure maximum compatibility and avoid interoperability issues.

- *Big Data Management:* The introduction of IoT solutions is closely tied to the work with high volumes of data. This is one of the challenges for telecom operators today that want to deliver IoT services. As the number of connected devices grows, the production of data also multiplies. Managing this data means having access to it, processing it in a timely manner, and storing an enormous amount of it. To deal with this task, they need to use agile and scalable computing platforms that would not negatively affect the productivity of apps.

Some of these challenges are depicted in Figure 5.9 [7].

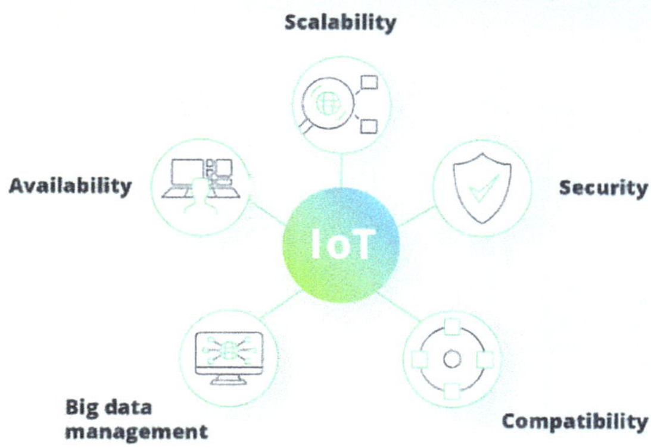

Figure 5.9 Some of these challenges of IoT in telecom [7].

5.6 CONCLUSION

The Internet of things (IoT) has progressed from a term to a solution with widespread use in homes and companies in recent years. It is expected to continue to change the roles of telecommunications service providers in enabling communication between people and devices. Integration of the IoT and telecommunication industry will become a solid competitive edge that optimizes productivity, improves service quality, unlocks extra opportunities, raises customer loyalty, increases profits, and leads the telecom business to success. Telecom companies are taking advantage of IoT technology to drive efficiency and optimize resource use within their own businesses. More information about IoT in the telecommunications industry is available from the books in [12-15].

REFERENCE

[1] "Telcos pitch trusted source norms for all network devices," June 2023,

https://www.financialexpress.com/business/industry-telcos-pitch-trusted-source-norms-for-all-network-devices-3136526/

[2] M. N. O. Sadiku, P. A. Adekunte, and J. O. Sadiku, "Internet of things in telecommunications," *International Journal of Trend in Scientific Research and Development*, vol. 8, no. 6, November-December 2024, pp. 253-262.

[3] M. Townes, "The spread of TCP/IP: How the Internet became the Internet,"

Millennium: Journal of International Studies, vol. 41, no. 1, 2012, pp. 43 –64.

[4] K. K. Patel and S. M. Patel, "Internet of things-IOT: Definition, characteristics, architecture, enabling technologies, application & future challenges," *International Journal of Engineering Science and Computing*, vol. 6, no. 5, May 2016, pp. 6122-6131.

[5] "Internet of things (IoT),"

https://www.geeksforgeeks.org/internet-things-iot-2/

[6] M.N.O. Sadiku, and S.M. Musa and S. R. Nelatury, "Internet of things: An introduction," *International Journal of Engineering Research and Advanced Technology*, vol. 2, no.3, March 2016, pp. 39-43.

[7] P. Sadeghi et al., "Towards a reliable modulation and encoding scheme for Internet of things communications," *13th IEEE International Conference on Application of Information and Communication Technologies (AICT)*, October 2019,

https://www.researchgate.net/figure/Communications-Technologies-in-Internet-of-Things_fig1_335104959

[8] R. Makarchuk, "The IoT in telecom: A data-driven path to

growth,"

https://intellias.com/the-iot-in-telecom-a-data-driven-path-to-growth/#:~:text=IoT%20initiatives%20provide%20telecommunications%20businesses,and%20generate%20better%20business%20ideas.

[9] "Internet of things in telecom: How your business gains revenue growth with IoT,"

https://startups.epam.com/blog/iot-in-telecommunications

[10] G. Lau, "The impact of IoT in telecom: How will it drive the future of this industry?" November 2022,

https://www.computer.org/publications/tech-news/trends/impact-of-iot

[11] E. Cole, "Telecommunications and the internet of things: Challenges, opportunities, and benefits" March 2022,

https://www.mytechmag.com/telecommunications-and-the-internet-of-things-challenges-opportunities-and-benefits/

[12] "IoT in telecom: Key use cases, benefits and challenges," December 2022,

https://www.cogniteq.com/blog/iot-telecom-how-its-used-and-what-benefits-it-provides

[13] T/ Kavitha et al. (eds.), Secure Communication in Internet of Things. Boca Raton, FL: CRC Press, 2024.

[14] M. O. Odhiambo and W. Mwashita (eds.), *Achieving Full Realization and Mitigating the Challenges of the Internet of Things (Advances in Wireless Technologies and Telecommunication)*. Engineering Science Reference, 2022.

[15] P. Kocovic, R. Behringer, and M. Ramachandran (eds.), *Emerging Trends and Applications of the Internet of Things (Advances in Wireless Technologies and Telecommunication)*. Information Science Reference, 2017.

[16] S. K. Routray and S. Mohanty (eds.), *Principles and*

Applications of Narrowband Internet of Things (NBIoT). Information Science Reference, 2021.

CHAPTER 6
CLOUD COMPUTING IN TELECOMMUNICATIONS

"The cloud services companies of all sizes...The cloud is for everyone. The cloud is a democracy."

—Marc Benioff

6.1 INTRODUCTION

Cloud computing is a means of pooling and sharing hardware and software resources on a massive scale. Users and businesses can access applications from anywhere in the world at any time. Companies offering these computing services are called cloud providers and typically charge for cloud computing services based on usage. Some features of cloud computing are displayed in Figure 6.1 [2]. The goal of cloud engineering is to create a highly scalable, flexible, and secure computing environment that can support the needs of modern businesses.

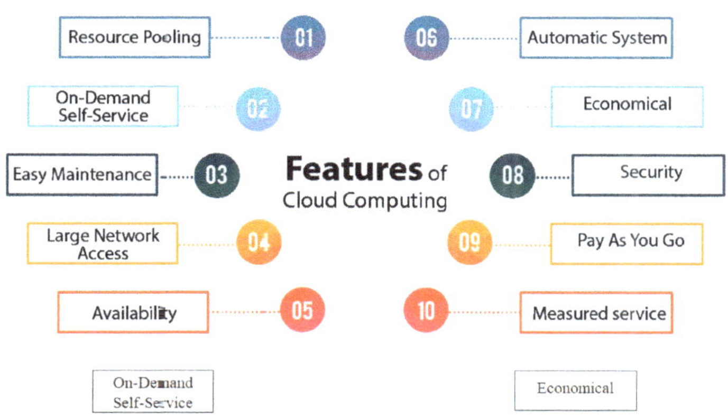

Figure 6.1 Some features of cloud computing [2].

The infrastructure that modern telecommunications providers maintain and operate is complex, as typically shown in Figure 6.2 [3].

Figure 6.2 The complexity of telecommunications infrastructure [3].

It integrates a vast portfolio of services interconnected with communication networks worldwide. These networks are wired together in complex ways. In the ever-evolving telecommunications industry, staying ahead of the competition is crucial for success. The telecommunication industry needs to provide distinctive service offerings beyond basic network connectivity. Cloud technology can help. Cloud computing has become a game-changer, empowering telcos to embrace agility, efficiency, and innovation like never before [4]. The telecom cloud market size growth is shown in Figure 6.3 [5].

Figure 6.3 The telecom cloud market size growth [5].

Cloud today is a domain where business and IT must work together. In spite of initial reservations due to migration complexities and security concerns, telecoms are now aggressively pursuing cloud-driven network transformations. Cloud computing in the telecom sector is an alternative to traditional telecom networks that increase agility and minimize operational costs. The role of cloud computing in the telecom industry is becoming increasingly prominent. The primary goal of cloud-enabled telecom networks is to manage networks effectively, minimize operational costs, and deliver new and better customer services. The cloud offers a scalable, flexible, and cost-effective platform that enables telecom operators to manage their network infrastructure, applications, and services efficiently [6].

This chapter examines the roles of cloud computing in the telecommunications industry. It begins with explaining what cloud computing is all about. It describes Telco cloud. It covers some applications of Telco cloud. It highlights the benefits and challenges of Telco cloud. The last section concludes with comments.

6.2 CLOUD COMPUTING BASICS

Cloud computing represents a newly emerging service-oriented computing technology. It is the provision of scalable computing resources as a service over the Internet. It allows manufacturers to use many forms of new production systems such as 3D printing, high performance computing (HPC), industrial Internet of things (110T), and industrial robots. It is transforming virtually every facet of modern manufacturing. It is innovating, reducing cost, and bolstering the competitiveness of American manufacturing [7].

The key characteristic of cloud computing is the virtualization of computing resources and services. Cloud computing is implemented in one of three major formats: software as a service (SAAS), platform as a service (PAAS), or infrastructure as a service (IAAS). These services are illustrated in Figure 6.4 [8] and explained as follows:

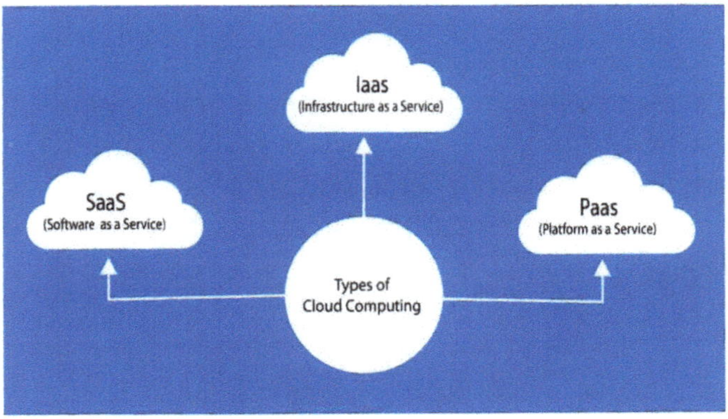

Figure 6.4 Three types of cloud computing services [8].

SaaS: This is a software delivery model in which software and associated data are hosted on the cloud. In this model, cloud service providers offer on-demand access to computing resources such as virtual machines and cloud storage. Nowadays oil & gas companies transition to cloud computing and implement SaaS solutions for operations.

PaaS allows the end-user to create a software solution using tools or libraries from the platform service provider. In this model, cloud service providers deliver computing platforms such as programming and execution.

In the laaS model, cloud service providers can rent manufacturing equipment such as 3D printers.

Just like cloud computing, CM services can be categorized into three major deployment models (public, private, and hybrid clouds) [9]:

- Private cloud refers to a centralized management effort in which manufacturing services are shared within one company' or its subsidiaries. A private cloud is often used exclusively by one organization, possibly with multiple business units.
- Public cloud realizes the key concept of sharing services with the general public. Public clouds are commonly implemented through data centers operated by providers such as Amazon, Google, IBM, and Microsoft.

- Hybrid cloud that spans multiple configurations and is a composed of two or more clouds (private, community or public), offering the benefits of multiple deployment modes.

These models are shown in Figure 6.5 [10]. Cloud computing finds application in almost every field.

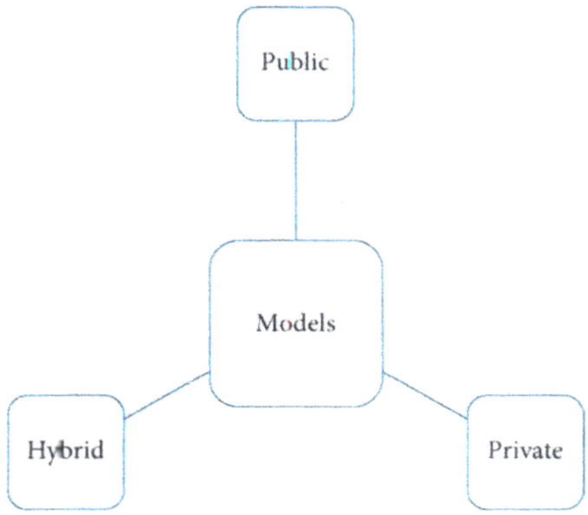

Figure 6.5 Cloud computing models [10].

6.3 TELCO CLOUD

Several telecom operators see value in trying to upgrade and simplify their networks. Telco cloud is a term used to describe the cloud-based services and infrastructure provided by telecommunications companies. It is a software-defined, highly resilient cloud infrastructure that allows telecommunications service providers (telcos) to add services more quickly, respond faster to changes in network demand, and manage central and decentralized resources more efficiently. It is mostly focused on creating a common virtualized infrastructure to enable the management ofvarious network functions required to deliver communication services. These cloud-based telecommunication services form a distributed computing network comprised of software-defined networking, network function virtualization, and cloud-native technology. The aim is to enable telcos to manage their networks, reduce costs, and deliver new services to their customers more effectively. Cloud-based solutions have revolutionized the IT field. They help businesses create a variety of services and applications that cannot easily be replicated by competitors [11].

The present and future of the teleo cloud is one that runs natively and efficiently in both multicloud and hybrid cloud environments. The primary business benefits of telco cloud are improved customer satisfaction, greater business agility, and cost reduction, as capital expenditures (CapEx) and operational expenditures (OpEx) are lowered through the use of non-specialized hardware and automation. Today, telcos use teleo cloud to create differentiation through new services and capabilities that cannot easily be replicated by competitors. Telco cloud makes it easy to co-create with enterprise service partners using public cloud services' anytime-anywhere-any device access. The telco cloud can extend to access resources outside of the operator's own private clouds to take advantage of public cloud capabilities and build a hybrid cloud environment [12]. Figure 6.6 shows a typical Telco cloud [13].

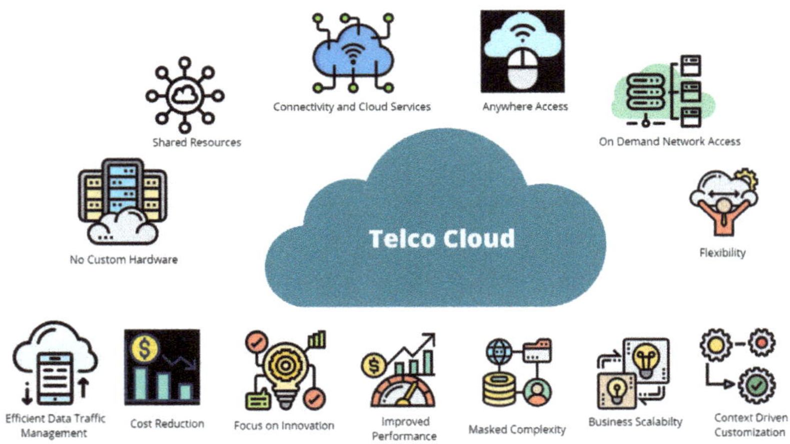

Figure 6.6 A typical Telco cloud [13].

6.4 APPLICATIONS OF TELCO CLOUD

Cloud computing has emerged as a transformative force within the telecommunications sector, offering a myriad of benefits that have reshaped the industry landscape. Cloud computing in the telecom industry refers to the utilization of virtualized infrastructure and software-driven solutions to deliver and manage telecommunications services. Telco cloud can be applied in the following ways [13]:

- *Global Interconnectivity:* Telecom operators have facilitated remote work and studying and helped people around the world stay connected. The demand for seamless global connectivity is rising. Telco cloud's ability to provide interconnectivity ross regions and networks ensures consistent services, making it a critical tailwind for international businesses and users. Now that ubiquitous connectivity commodity, the telecommunications companies (or telcos) need to provide distinctive service offerings beyond basic network connectivity. Cloud technology can help.

- *Cloud Services:* Cloud computing is enabling telcos to offer services with unmatched agility. Telco cloud solutions help businesses create a variety of services and applications that cannot easily be replicated by competitors. Cloud computing allows telecom service providers to offer a broader range of services, including voice, data, and multimedia communication, with greater flexibility and efficiency. Telecom cloud service is clearly the path forward to transform operations, strengthen competitiveness and spearhead digital innovation. Embracing cloud technologies enables telecom companies to modernize their infrastructure, streamline operations, and deliver enhanced services to their customers. Figure 6.7 shows a field engineer [14].

Figure 6.7 A field engineer [14].

- *5G Services:* The fifth generation of mobile networks, 5G, offers markedly higher speeds, lower latency, and the capability to facilitate more devices than ever before. In addition to traditional telecom services, cloud is particularly valuable for 5G deployment and offering new services, such as Internet of things (IOT), machine-to-machine communication, and robotics. 5G will be a game changer that strengthens telecom network infrastructure. Telecom providers are increasingly adopting multicloud strategies to deliver 5G services efficiently and meet evolving consumer needs. The evolution of 5G networks brings exciting trends like edge computing, network slicing, IOT, artificial intelligence (Al), and cloud services. Al integrated into telecom cloud platforms significantly enhances these capabilities.

- *Edge Computing Integration:* Telco cloud, integrated with edge computing capabilities, enables quicker response times and enhanced user experiences, a crucial tailwind in the era of instant gratification. In response to the increasing demand for low-latency services, such as the Internet of things (IoT) and 5G, telecom companies are strategically integrating edge computing into their cloud infrastructure. Edge computing enables data processing and analysis to occur closer to the source of data generation, reducing latency and improving the overall performance of latency-sensitive applications. The Internet of things, ubiquitous globally, is based on edge computing. For telecom, reducing latency is critical in terms of customer service.

- *Al and ML Integration:* Artificial intelligence (Al) and machine learning (ML) are becoming integral parts of telecommunications operations. Telco cloud offers the computational power needed for Al and ML algorithms, driving intelligent automation, predictive analytics, and enhanced customer services. Generative Al (GenAI) is crucial in streamlining customer services, enabling communication service providers (CSPs) to manage communication traffic efficiently and bolster security. It is becoming increasingly clear that every telecom company must realize the necessity of being Al-driven.

6.5 BENEFITS

Building a telco cloud on open-source software brings a wide range of benefits for your business. Telecommunications cloud solutions unlock a variety of new opportunities and hence revenue streams for telecom operators. They offer multiple benefits, such as improved scalability, cost efficiency, global reach, etc. They open up new opportunities to create and deliver business value as well as increase profitability. Other benefits of Telco cloud include [11,15]:

- *Customer Satisfaction:* Cloud computing might turn out to be an advantageous go-the-automation route in sustained customer engagement. Telco cloud infrastructure brings improved service resilience and the ability to quickly respond to faults and fluctuations in demand. This allows operators to maintain service levels and competitive pricing, which inevitably results in reduced customer churn.

- *Scale Elastically:* Telecom cloud extends logically across data centers, central offices, the public cloud, and edge locations, creating almost limitless capacity from a hybrid environment. Businesses are empowered to quickly move workloads to handle traffic spikes, work around faults, accommodate additional remote workers, onboard new customers, spin applications up and down when needed, etc.

- *Cost Reduction:* One of the greatest advantages that cloud has to offer is the substantial IT cost savings. Cloud computing helps telecom companies save costs in multiple ways. Cloud infrastructures are innately cost effective due to their highly flexible, pay-as-you-go models. By using a network of remote servers, businesses are able to rent and pay for exactly the resources they need and significantly reduce overhead expenses for things like software and hardware updates, backup systems, and data storage. Cloud computing is beneficial for telecom companies as it allows them to use their resources more efficiently by allocating them based on demand. Cloud computing enables telecom companies to reduce costs, improve operational efficiency, and allocate resources more effectively, leading to substantial savings in the long run.

For example, Vodafone migrated its IT infrastructure to the cloud, resulting in significant cost savings and increased agility.

• *Scalability:* The ability to scale resources up or down based on demand is a gamechanger for telcos. Cloud scalability means the ability to stay flexible and adapt to fluctuating storage and infrastructure needs by automatically provisioning and deprovisioning resources. Cloud computing for telecom enables businesses to strategically upscale and downscale their operations to maximize and capitalize on growth opportunities. With high scalability, telecommunications service providers can effectively reduce costs without compromising resources.

• *Flexibility:* Cloud computing is becoming an important player in the telecom industry, as it offers scalability and flexibility to meet the increasing demands for data, bandwidth, and innovative services. Cloud computing enables telcos to handle fluctuating customer requirements, adapt to changing market conditions, and stay flexible in an ever-changing landscape.

• *Market Adaptability:* Driven by shifting customer demands, market conditions are constantly changing too. The ability to respond quickly to varying consumer needs becomes a key differentiator for businesses. The cloud also supports innovation and enables operators to easily revamp and update their telecommunications cloud services.

• *User-Friendliness:* Cloud computing is highly convenient and user-friendly, as it places minimal load on businesses. Cloud network functions deployment, faster critical solutions, and quick response to customer requests benefit the telecom market. Telecom companies use cloud platforms to provide self-service options for customers, allowing them to manage their accounts, access services, and troubleshoot issues independently. Telecommunications and cloud computing work together to generate environmentally friendly solutions.

• *Business Agility:* Cloud computing enables telcos to deliver services with unmatched agility. Through virtualization and automation, telcos can rapidly' deploy and scale network services, reducing time-to-market for new offerings. This agility allows telcos to respond swiftly to changing market demands, ensuring they seize new business opportunities promptly,

- *Enhanced Customer Experience:* Cloud-based customer relationship management (CRM) and customer experience platforms empower telcos to personalize services, anticipate customer needs, and deliver a superior user experience. By harnessing cloud-powered analytics and AI capabilities, telcos gain valuable insights into customer behavior, enabling data-driven decisions to enhance service offerings.

- *Affordability:* Cloud computing services often come with lower costs and greater affordability as users only pay for the resources they consume, eliminating the need for upfront investments in hardware and infrastructure. Figure 6.8 shows some main benefits of Telco cloud [5].

Figure 6.8 Main benefits of Telco cloud [5].

6.6 CHALLENGES

Telecom companies are generally satisfied with the results delivered b oud but also face challenges when it comes to cloud usage, competition, and cost. It is not that easy to shift telco's infrastructure and its applications to the cloud. Other challenges of Telco cloud include the following [14]:

- *Security:* The more data that is stored and processed in the cloud, the greater the security risks. With the increasing adoption of cloud computing in telecommunications, there is a growing focus on security and compliance to protect sensitive data and ensure regulatory compliance. Reputable cloud service providers maintain robust data centers with stringent security measures. Leveraging cloud infrastructure ensures data security and business continuity, providing telcos with peace of mind as they store and process sensitive customer information. Sophisticated cloud systems can open up new cybersecurity challenges. Cloud security solutions such as encryption, access control, and threat detection help mitigate risks and enhance the resilience of telecom cloud environments against cyber attacks.

- *Privacy:* Ensuring data privacy, protecting against cyber threats, and complying with regulations remain significant challenges for Telco cloud providers. Cloud computing providers are now highly concerned about their users' data security, sensitive data storing, and privacy. User data is made safer by avoiding storage on a single device and by using encrypted passwords and restricted access.

- *Holistic Approach:* When migrating to a cloud-native architecture, a holistic approach is key. The migration might be implemented one network function or service at a time, but the process should start with a comprehensive cloud readiness assessment that encompasses infrastructure, applications and service portfolios, organization, and processes.

- *Legacy Infrastructure:* Many telecom companies still rely on legacy systems, which require a pre-update before migrating to the cloud. Integrating these systems with advanced Telco cloud technologies can be challenging and costly, acting as a headwind in the path of seamless migration. Migration to a cloud infrastructure is a complex process that requires certain efforts and costs.

- *Regulation:* Telecommunications is a heavily regulated industry, and compliance with various regulatory frameworks globally can be complex. Navigating these regulations, especially concerning data storage and privacy, can pose challenges for Telco cloud providers.

- *Complexity:* The more coverage and the number of users a

telecom operator has, the more complex and chaotic the operational processes become. Requests and tasks grow like a snowball: network maintenance, accounting and reporting, making payments, etc. All this requires human resources and efficient processing tools.

- *Personalized Services:* Personalization is an integral part of service in any industry. Users are used to getting individual offers and a quality digital experience. Telecom is no exception. That is why telecom operators are forced to constantly improve the user experience, create new offers to attract the audience better and meet high support standards.

6.7 CONCLUSION

Cloud computing is a key element of the modernization of telecom networks and services. It is the practice of using the capacities of remote servers and their networks to collect, store, manage, and process data. Cloud computing in telecom is a software defined infrastructure that allows telcos to store and process data remotely in the data processing centers. Looking ahead, the cloud's potential is vast. The continued development of technologies like Al and machine learning within the cloud environment will further enhance network management, automate tasks, and personalize customer experiences.

As telcos continue to embrace cloud technologies, they position themselves to thrive in a rapidly evolving digital landscape, offering cutting-edge services and responding swiftly to the ever-changing demands of modern communication. Cloud-based communication solution have become a central force in various industries worldwide. Cloud-based solutions will play a crucial role in enabling next-generation technologies like 6G and the Internet of things (IoT). More information about cloud computing in the telecommunications industry can be found in the books in [16,17].

REFERENCE

[1] M. N. O. Sadiku, S. M. Musa, and O. D. Momoh, "Cloud computing: Opportunities and challenges," *IEEE Potentials*, January-February 2014, pp. 34-36.

[2] "Features of cloud computing — 10 major characteristics of cloud computing," https://data-flair.training/blogs/features-of-cloud-computing/

[3] "Cloud computing in telecom: Sats, benefits, best practices," May 2021, https://www.n-ix.com/cloud-computing-telecom/

[4]"Cloud computing applications and benefits in telecom," August 2023, https://www.linkedin.com/pulse/cloud-eomputing-applications-benefits-telecom-systemsmea/

[5]"Cloud computing for telecom industry: Role, benefits and ways of deployment," August 2021, https://www.aimprosoft.com/blog/cloud-computing-and-telecommunications/

[6] M. N. O. Sadiku, C. M. M. Kotteti, and J. O. Sadiku, "Cloud computing in telecommunications industry," *International Journal of Trend in Research and Development*, vol. 11, no. 6, 2024, pp. 57-61.

[7] S. Ezell and B. Swanson, "How cloud computing enables modern manufacturing", June 2017,

https ://itif.org/publications/2017/06/22/how-cloud-computing-enables-modernmanufacturing

[8] "Cloud computing applications in agriculture," https://www.eescorporation.com/cloud-computing-applications-in-agriculture/

[9] "Cloud manufacturing," Wikipedia, the free encyclopedia https://en.wikipedia.org/wiki/€loud_manufacturing

[10] J. S. Saini et al., "[Retracted] cloud computing: Legal issues and provision," Security and Communication Networks, vol. 2022,

https://www.hi dawi.com/jo rnals/scn/2()22/2288961/

[11] "Cloud technology in telecom Industry: Key revenue streams and opportunities,"

https://techsphere-distribution.com/blogs/insights/cloud-technology-in-telecom-industrynew-revenue-streams-and-opportunities

[I 2] "What is telco cloud?" February 2023,

https://www.redhat.com/en/topics/cloud-computing/what-is-telco-cloud#:~:text=Telco%20cloud%20is%20a%20software.and%20decentralized%20resources%20more%20efficiently.

[13] A. Puglia, "Telco cloud evolutions," October 2023, https://mwv.linkedin.com/pulse/telco-cloud-evolutions-alessandro-puglia/

[14] A. Solovei, "Cloud computing for the telecom industry," November 2022

https://ardas-it.com/cloud-computing-for-the-telecom-industry

[15] "How is cloud computing helping the telecom industry grow?" April 2024,

https://www.travancoreanalytics.com/cloud-computing-in-telecom-industry/#:~:text=Overall%2C%20cloud%20computing%20enables%20telecom.savings%20in%20the%20long%20run.

[16] X. Zhang et al. (eds.), *Cloud Computing, Smart Grid and Innovative Frontiers in Telecommunications*. Springer, 2020.

[17] J. Rodrigues and K. Lin (eds), Mobile Networks and Cloud Computing Convergence for Progressive Services and Applications. IGI Global, 2013.

CHAPTER 7
EDGE COMPUTING IN TELECOMMUNICATIONS

"Any sufficiently advanced technology is equivalent to magic."

—Arthur C. Clarke

7.1 INTRODUCTION

The proliferation of Internet of things (IoT), the success of cloud services, and 5G communication technologies have led to the emergence of a new computing paradigm, edge computing, which calls for processing the data at the edge of the network, as opposed to a data center or cloud. This is part of a larger cloud computing architecture that requires that computational needs can be met "at the edge," where the data is collected. Existing cloud computing has not been able to meet the requirements of low latency, location awareness, and mobility support [1]. Cloud computing is not an efficient way to process data which is produced at the edge of the network.

Telecommunications networks are vast and extremely distributed, with critical network infrastructure deployed at core sites like Internet exchanges and data centers, business and residential customer premises, and access sites like towers, street cabinets, and cell site shelters. The infrastructure that modern telecommunications providers maintain and operate is complex, as typically shown in Figure 7.1 [2].

Figure 7.1 The complexity of telecommunications infrastructure [2].

In the ever-evolving landscape of telecommunications, the emergence of edge computing stands as a beacon of promise, offering ultra-low latency solutions that propel networks into the future. By processing data locally at the edge, telecom operators can deliver ultra-low latency experiences, critical for applications like real-time video analysis, online gaming, and autonomous vehicles. This near-instantaneous responsiveness enhances user satisfaction.

Edge computing represents a shift from traditional centralized data processing to a distributed architecture, allowing for data processing at the network edge, closer to the devices and sensors generating the data. This approach offers reduced latency, improved scalability, and enhanced security. Edge computing is a strategic opportunity for telcos. Its most significant advantage in telecom is the drastic reduction in latency. Adopting edge computing is a high priority for many telecommunications service providers as they modernize their networks. The integration of edge computing into telecom networks represents a paradigm shift that promises to revolutionize the way we connect, communicate, and collaborate in the digital age. Edge computing is poised to shape the future of telecommunications and usher in a new era of connectivity and innovation [3].

In this chapter, we explore the evolution and applications of edge computing in the telecommunications industry. The chapter begins with explaining the concept of edge computing. It describes the Telco edge computing. It provides some applications of Telco edge computing. It highlights the benefits and challenges of Telco edge computing. The last section concludes with comments.

7.2 CONCEPT OF EDGE COMPUTING

The history of edge computing traces back to the introduction of content delivery networks in the 1990s. The concept was straightforward: place servers close to end-user locations for faster cached image and video transmission. Edge computing was created jointly by Microsoft and their academic collaborators. Today, more and more services are pushed from the cloud to the edge of the network. Since data is increasingly being produced at the edge of the network, it would be efficient to process the data at the edge of the network. Keeping this data closer to its users (at the edge) eliminates many of the problems inherent with the public cloud model. Figure 7.2 shows how edge computing works [2]. Edge computing allows data from IoT devices to be analyzed at the edge before being sent to the cloud.

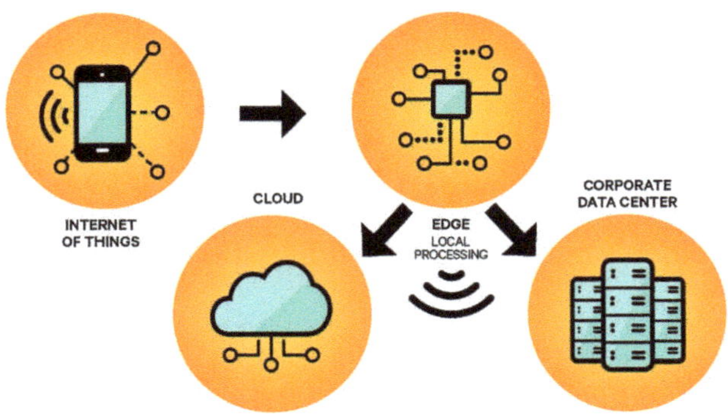

Figure 7.2 How edge computing works [2].

Edge computing (EC) or edge cloud is a computing paradigm where substantial compute and storage resources are placed at the edge of the Internet, in close physical proximity to mobile devices, sensors, end users, and IoT devices. It refers to bringing the flexibility and openness of cloud-native infrastructure to that local infrastructure. The idea of "edge" is to do processing near the data source. The terms "cloudlets," "micro data centers," and "fog" are used to refer to these small, edge-located computing nodes or data centers. A cloudlet is a cluster of computers well connected to the Internet and can be treated as" data center in a box." The main objective of a cloudlet is to extend the remote datacenter cloud services in close proximity to the end users. Physical proximity is the essence of edge computing since it improves latency, bandwidth, trust, and survivability. While the cloud has revolutionized the way we deal with data, the next wave of that revolution will happen at the edge [4].

A standard edge computing framework consists of three distinct levels [5]:

- *The cloud:* Which manages the overall data storage and processing.
- *The edge:* Tasked with near-instantaneous data handling.
- *The device:* Responsible for initial detection and basic data processing.

These three levels are related in the edge computing architecture shown in Figure 7.3 [6].

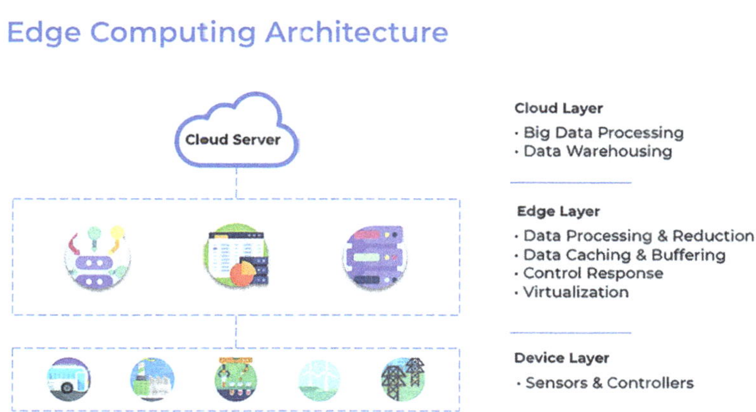

Figure 7.3 Edge computing architecture [6].

Edge computing covers a wide range of technologies such as wireless sensor networks, distributed data storage, and augmented reality. This has made its way to becoming the core of the data center. The term "edge" refers to the computing devices that sit closer to the sources of data, where the digital world meets the real world. These edge devices typically reside away from the centralized computing available in the cloud and are being created with increasingly compute capabilities. Typical edge devices are smartphones, tablets, sensors, wearables, routers, switches, integrated access devices, multiplexers, smart TV, modern cars, and a variety of MAN/WAN access devices. Edge computing enables analytics and data gathering to occur at the source of the data.

Edge computing covers a spectrum of technologies such as cloudlets, fog computing, and mobile edge computing. A combination of edge and cloud computing is referred to as fog computing because it combines centralized and distributed computing resources into a single architecture. (Edge computing is a relatively new concept that should not be confused with fog computing.) It is practically unsafe and unnecessary to send such a large amount of data to the cloud. A comparison between cloud computing, edge computing, and fog computing is made in Table 4.1 [7] and illustrated in Figure 7.4 [8].

Characteristics	**Cloud**	**Fog**	**Multi-cloud**	**Edge**
Latency	High	Low	Very High	Low
Bandwidth Utilization	High	Low	Very High	Very Low
Response Time	High	Low	High	Low
Storage	High	Low	Very High	Low
Server Overhead	Very High	Low	High	Very Low
Energy Consumption	High	Low	High	Low
Network Congestion	Very High	Low	High	Low
Scalability	Medium	High	Medium	High
Quality of Service and Quality of Experience	Medium	High	Medium	High

Figure 7.1 A comparison of cloud, edge, and fog computing [7].

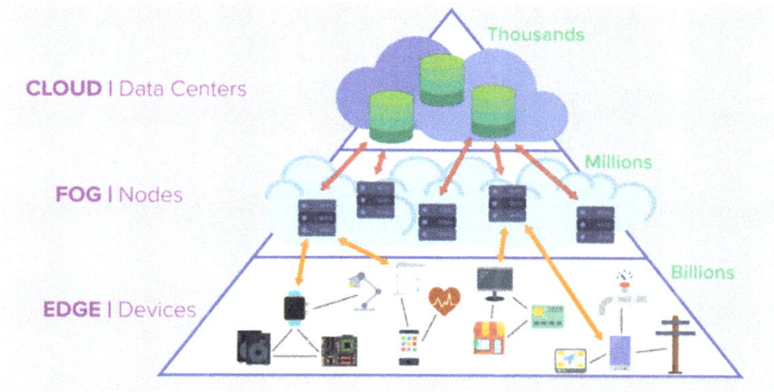

Figure 7.4 The relationship between cloud computing, edge computing, and fog computing [8].

Compared to cloud computing, fog computing and edge computing have the following five advantages [9,10]: (1) greater data transmission speed, (2) less dependence on limited bandwidths, (3) greater privacy and security, (4) greater control over data generated in foreign countries where laws may limit use or permit unwanted governmental access, and (5) lower costs because more sensor-derived data are used locally, and less data are transmitted remotely.

7.3 TELCO EDGE COMPUTING

In traditional telecom networks, data is typically processed in centralized data centers located far away from the end users. This can result in latency issues, as data has to travel long distances before reaching its destination. Edge computing solves this problem by bringing data processing closer to the edge of the network, where it is generated.

Telco edge cloud is mainly defined as distributed compute managed by a telco. This includes running workloads on customer premises equipment (CPE) at customers' sites as well as locations within the operator network such as base stations, central offices, and other aggregation points on access and/or core network. Edge devices preprocess and filter data locally, transmitting only relevant information to the cloud. The telecom edge is the catalyst creating a new world where the telecom and cloud industries join forces to eliminate duplication while creating a new era of latency-sensitive applications and services. Figure 7.5 shows edge computing use cases in telecom [5].

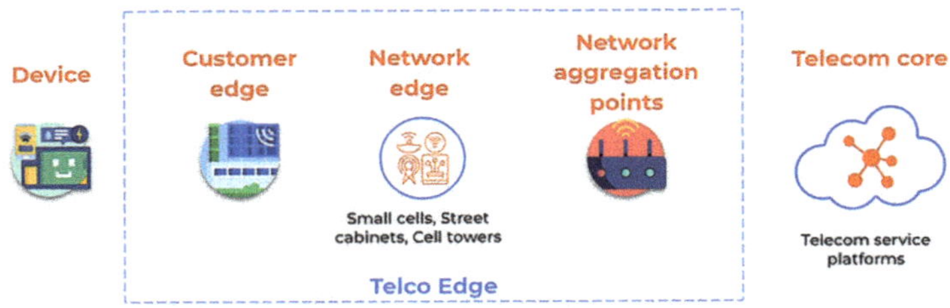

Figure 7.5 Edge computing use cases in telecom [5].

7.4 APPLICATIONS OF TELCO EDGE COMPUTING

Edge computing is the concept of capturing and processing data as close to its source or end user as possible. It is a distributed computing architecture framework that brings computer processing from a distant data center closer to the source of data generation and consumption. This is necessary to reduce latency by moving away from completely centralized operations. Edge computing is revolutionizing the telecom industry by bringing computing power closer to the source of data generation. By leveraging the power of edge computing, telecom operators can deliver unparalleled connectivity, drive digital transformation, and unlock new revenue streams. Some use cases of Telcom edge as displayed in Figure 7.6 [5]. Telcom edge computing is applied in telecom in the following ways [6,11,12]:

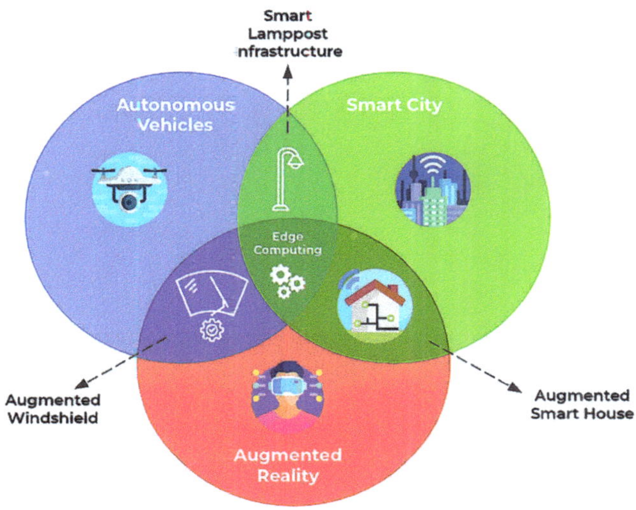

Figure 7.6 Some applications of Telcom edge [5].

- *Telecom Edge Computing and 5G:* Telecommunications providers are rolling out 5G networks around the globe. These new cellular telecommunications networks provide exponentially more bandwidth and faster processing speeds than today's 4G networks. Telecom operators envisioned edge computing as a way to monetize 5G to deliver big dividends. Extensive 5G deployments, for example, could set the stage for momentum in the edge computing space. Discussions on edge computing strategies and market are often linked to 5G. The combination of 5G and edge computing has the potential to be a catalyst for innovation across numerous industries. 5G improves speed by increasing spectral efficacy and offers the potential of much higher speeds than 4G, while edge computing reduces latency by shortening the time required for data processing by allocating resources closer to the application. The combination of 5G's higher connectivity speeds and edge computing has the potential to be a catalyst for innovation across numerous industries. It can support advanced robotics, artificial intelligence (AI) and autonomous technology, massive Internet of things (IoT), and augmented and assisted reality (AR). Figure 7.7 shows the combination of 5G and edge computing [4].

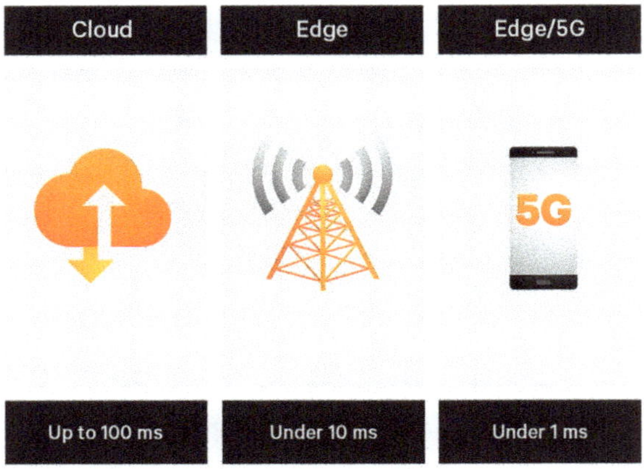

Figure 7.7 The combination of 5G and edge computing [4].

- *Telecom Edge Computing and IoT:* Edge computing plays a crucial role in enabling real-time data processing for IoT applications. It is a game-changer for IoT applications. The telecommunications industry is one of many that can benefit from IoT technology, devices, and sensors. IoT is particularly useful (and cost-effective) for equipment in remote or hard-to-reach locations. This data can also be used for preventative maintenance, to predict and mitigate faults in networks and equipment before they even occur, which can further reduce costs.

- *Telecom Edge Computing for Autonomous Vehicles:* Autonomous vehicles are one of the most promising applications of edge computing in the telecom industry. Edge computing is revolutionizing the capabilities of autonomous vehicles in the telecom industry. By enabling real-time data processing, reducing latency, and enhancing security, edge computing plays a vital role in ensuring the safe and efficient operation of autonomous vehicles. This eliminates the reliance on distant cloud connections, ensuring that autonomous vehicles can operate efficiently and safely. One of the key advantages of edge computing for autonomous vehicles is the ability to handle large amounts of data in real-time. The sensors on autonomous vehicles generate massive amounts of data that need to be processed quickly to enable safe navigation.

- *Edge Computing for AR/VR:* Augmented reality (AR) and virtual reality (VR) are rapidly growing technologies that are revolutionizing various industries. These immersive technologies require real-time processing of large amounts of data to create seamless and interactive experiences. However, the latency and bandwidth limitations of traditional cloud computing models can hinder the performance of AR and VR applications. Even minor latency in AR/VR can disrupt the experience and even induce discomfort or visual sickness. Low latency guarantees that there is as little delay between what an individual does and the system's response. This is where edge computing comes in. Edge computing enables data processing at the network edge, closer to the devices and sensors used in AR and VR applications. By bringing the processing power closer to the end-users, edge computing reduces latency and ensures a smooth and immersive experience. Augmented and virtual reality reduces response time and improves customer experience.

- *Video Surveillance:* Using video to identify key events is rapidly spreading across all domains and industries. Transmitting all the data to the cloud or data center is expensive and slow.. With edge computing, for example, cameras that are located close to the event can determine whether a human is caught in the fire by identifying characteristics typical of a human being and clothing that humans might normally wear which might survive the fire. As soon as the camera recognizes a human in the video content, it will start transmitting the video to the local edge.

- *Smart Cities:* Cities, school campuses, stadiums, and shopping malls are examples of the many places where edge AI is transforming locations into smart spaces. Operating and governing cities has become a challenging mission due to many interrelated issues such as increasing operation cost from aging infrastructure, operational inefficiencies, and increasing expectations from a city's citizens. IoT devices are the basic building blocks of any smart city solution. Embedding these devices into the city's infrastructure and assets helps monitor infrastructure performance and provides insightful information about the behavior of these assets. Edge computing becomes a necessary technology in order to deliver city's mission critical solutions such as traffic, flood, security and safety, and critical infrastructure monitoring.

- *Manufacturing:* Rapid response to manufacturing processes is essential to reduce product defects and improve efficiencies. Different edge devices are capable of differing levels of processing and relevant information can be sent to multiple edges including the cloud. Operating at peak efficiency and with no unplanned outages is the difference between having profit and not having profit. Industrial edge and cloud computing are among the most exciting areas of technology development for smart manufacturing.

7.5 BENEFITS

Edge computing is revolutionizing the telecom industry by enabling real-time data processing, enhancing network performance, improving data security, and reducing costs. One of the most significant advantages of edge computing in telecom is the drastic reduction in latency. Another key benefit of is its ability to support the growing demand for high-bandwidth applications and services. Other benefits include [12]:

- *Latency:* Excessive delay in telecom services can significantly impair the quality of user interactions. The modern digital world and its products demand low latency. One of the primary benefits of edge computing is that it greatly reduces the effects of latency on applications. This enables new applications and services on the network that can exploit reduced latency and improve the experience of existing apps. When a service provider moves mobile workloads closer to the end user, increasing throughput and reducing latency, it can be considered a new kind of mobile architecture. Figure 7.8 displays the impact of high latency on telecom services [5].

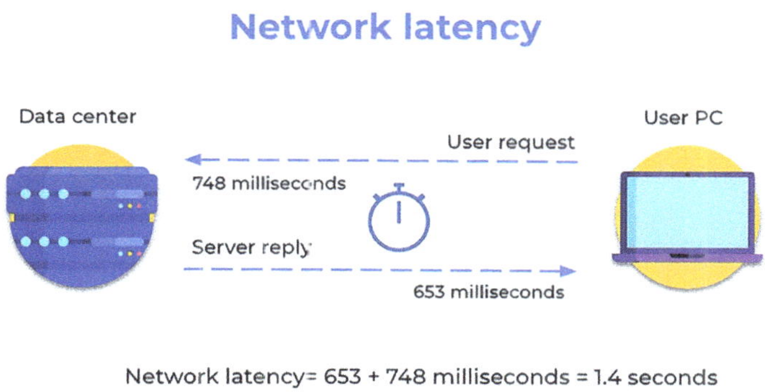

Figure 7.8 The impact of high latency on telecom services [5].

- *Reduced Bandwidth:* When data is sent to the cloud, it travels through a wide area network, which can be costly due to its global coverage and high bandwidth needs. When processing data at the edge, local area networks can be utilized, resulting in higher bandwidth at lower costs. Edge computing alleviates bandwidth bottlenecks and minimizes service disruptions.

- *Revenue Generation:* Edge computing in telecom offers a range of opportunities for monetization and revenue generation. For telcos, the apps and services their customers want to consume on edge networks are the key to revenue generation. Telco providers in particular have the credibility, skills, and relationships to capture these revenue-generating opportunities in the edge market. Telecom edge computing offers operators a strategic opportunity to enhance operations and leverage their technology expertise to generate more revenue

- *Efficiency:* Telecommunications companies can use edge computing in a number of ways to improve the efficiency of their operations. Edge computing can contribute to energy efficiency by reducing the need for data transfer and optimizing resource utilization. This can lead to cost savings in energy consumption.

- *Enhanced Network Performance:* Enhanced network performance and latency reduction are crucial factors in the telecom industry. These factors are necessary in providing seamless connectivity and delivering high-performance, low-latency, and secure services to end-users. One of the key technologies that enable enhanced network performance and latency reduction is edge computing. This approach also offers enhanced security and the ability to operate offline.

- *Availability:* Critical systems need to operate irrespective of connectivity. There are many potential points of failure with the current communications flow. With edge computing, with the communication primarily between the consumer/requestor and the device/local edge node, there is a resulting increase in availability of the systems.

- *Data Security:* In edge computing architectures, the analytics data potentially never leaves the physical area where it is gathered and is used within the local edge. Only the edge nodes need to be primarily secured, which makes it easier to manage and monitor, which means that the data is more secure.

- *Data Sovereignty:* When data is processed at the location it is collected, edge computing allows organizations to keep all of their data and compute at a suitable location. This results in reduced exposure to cybersecurity attacks and adherence to strict, ever-changing data location laws.

7.6 CHALLENGES

Although edge computing opens up many prospects for telecom, it also presents some challenges that telcos must address. Challenges of edge computing in the telecom industry include network connectivity and reliability, data management and processing, and standardization and interoperability. Other challenges include the following [12]:

- *Privacy and Security:* Data security and privacy are of utmost importance in telecommunications. While data security is a benefit in that the data can be limited to certain physical locations for specific applications, overall security is an additional challenge when adopting edge computing. Device edge physical devices might not have the ability to leverage existing security standards or solutions due to their limited capabilities. With the increasing amount of sensitive information being transmitted over networks, it is crucial to ensure that this data is protected from unauthorized access. Edge computing addresses these challenges by keeping sensitive data localized, reducing the risk of exposure during transit. By processing data closer to its source, telecom operators can implement robust security measures at the edge, safeguarding sensitive information and ensuring compliance with regulatory requirements. Potential edge computing security issues also hamper widespread adoption.

- *Reliability:* Network connectivity and reliability are crucial factors in the success of edge computing in the telecom industry. Reliable network connectivity ensures that data can be transmitted efficiently and without interruption, enabling real-time data processing and analysis. Without reliable network connectivity, edge computing applications may experience latency issues and data loss, compromising the overall performance and effectiveness of the system. To address the challenges of network connectivity and reliability, telecom companies are investing in diverse connectivity technologies such as Wi-Fi 6.

- *Modernization:* Telco providers face complex challenges that put the impetus behind modernizing their networks. These include simplifying network operations, improving flexibility, availability, efficiency, reliance, and scalability—while reducing latency and providing better application response times by processing and storing data closer to users and devices.

- *Data Management:* This becomes critical in edge computing, as there is a need to balance local processing and storage with cloud-based operations. Ensuring security and privacy at the edge is essential, as edge devices are more susceptible to physical and cyber threats. However, these challenges also present opportunities for innovation, such as the development of edge AI algorithms and optimization techniques tailored for resource-constrained environments.

- *Standardization and Interoperability:* These are key challenges in the adoption of edge computing in the telecom industry. With many standards in this ecosystem newly created or quickly evolving, it will be difficult to maintain technology decisions long term. As edge computing involves the deployment of distributed infrastructure and the integration of various technologies, it is essential to establish common standards and protocols to ensure seamless interoperability between different edge computing systems. Without standardized interfaces and protocols, the integration and coordination of edge computing systems can be complex and time-consuming, hindering the scalability and efficiency of edge deployments. Edge-based platforms also help smooth application development operations, offer better interoperability, and support a wide range of hardware and software environments.

- *Compatibility:* Compatibility issues have hindered edge computing adoption. Edge computing typically requires hardware and software from multiple vendors. The lack of standardization and subsequent interoperability and integration challenges have also hindered adoption and success.

- *Designing Infrastructure:* Designing the infrastructure for an edge application can be challenging. Compute needs, physical environment, power and thermal management requirements, physical security, and remote monitoring and management capabilities must all be considered. Edge standards and devices are changing rapidly. Making the wrong decision could mean that you could need to re-invest in IT sooner than expected.

7.7 CONCLUSION

Although edge computing is still evolving computing and has not lived up to the hype, it holds much promise for the future of telecommunications. Edge computing has undeniably revolutionized the telecom sector, minimizing latency for enhanced user interactions by bringing data processing closer to the source of data generation. It enables real-time data processing and analysis, improving network performance and reducing latency. As telecom operators navigate the complexities of a rapidly evolving landscape, embracing edge computing is essential to meet the evolving needs of consumers and enterprises alike.

Edge computing is a game-changer for the telecom industry, offering a range of benefits including improved performance, reduced latency, enhanced security, and new revenue opportunities. The emergence of edge computing stands as a beacon of promise, offering ultra-low latency solutions that propel networks into the future. In order to truly reap the benefits that edge computing promises, telco providers need to carefully build their strategy and develop the necessary technologies and skills internally. More information about edge computing in the telecommunications industry can be found in the book in [13].

REFERENCE

[1] E. Ahmed and M. H. Rehmani, "Mobile edge computing: Opportunities, solutions, and challenges," *Future Generation Computer Systems*, vol. 70, 2017, pp. 59-63.

[2] "Harnessing edge computing for business growth,"

https://www.vertiv.com/en-us/solutions/vertiv-guide-to-edge-computing/

[3] M. N. O. Sadiku, C. M. M. Kotteti, and J. O. Sadiku, "Edge computing in telecommunications industry," *International Journal of Science, Engineering and Technology*, vol. 12, no. 6, 2024, pp. 1-8.

[4] M. N. O. Sadiku, A. M. Oteniya, and S. M. Musa, "Edge computing," *International Journal of Trend in Research and Development*, vol. 5, no. 4, July – Aug. 2018, pp. 553-554.

[5] A. Harper, "How is edge computing enhancing telecom speeds for end users?"

https://www.jevera.software/post/how-is-edge-computing-enhancing-telecom-speeds-for-end-users

[6] J. Gonzalez et al., "Edge computing architecture and use cases,"

https://lfedge.org/edge-computing-architecture-and-use-cases/

[7] H. El-Sayed et al., "Edge of things: The big picture on the integration of edge, IoT and the cloud in a distributed computing environment," *IEEE Access*, vol. 6, 2018, pp. 1706-1717.

[8] "Edge computing needs edge AI," August 2018,

https://www.imagimob.com/blog/edge-computing-needs-edge-ai#

[9] Y. Song, "An Approach to QoS-based Task Distribution in Edge Computing Networks," *Doctoral Dissertation, Arizona State University*, December 2018.

[10] D. C. Klonoff, "Fog computing and edge computing

architectures for processing data from diabetes devices connected to the medical Internet of things," *Journal of Diabetes Science and Technology*, vol. 11, no. 4, 2017, pp. 647 –652.

[11] A. Jadhav, "Cisco, GDIT to deliver private 5G networks for public sector edge computing," July 2022,

https://www.edgeir.com/cisco-gdit-to-deliver-private-5g-networks-for-public-sector-edge-computing-20220706

[12] "The impact of edge computing in the telecom industry," December 2023,

https://metavshn.com/the-impact-of-edge-computing-in-the-telecom-industry/

[13] A. Yarali, *Cloud, Fog, and Edge: Technologies and Trends in Telecommunications Industry*. Nova Science Pub Inc, 2018.

CHAPTER 8
BLOCKCHAIN IN TELECOMMUNICATIONS

"The blockchain is a distributed network that solves all the problems that we have of finance, but more broadly, it's like a philosophy. It's a way of life."

—Mike Cernovich

8.1 INTRODUCTION

Recently, modern technologies have aided business titans in deciding to use digital currency rather than conventional trade in order to avoid fraud. Blockchain is a shared ledger that is replicated across users in a particular network. It enables the existence of Bitcoin and many other cryptocurrencies. We all associate blockchain with Bitcoins. A cryptocurrency refers to a digital asset that works as a medium of exchange between various business organizations. Blockchain technology has gained significant popularity in various industries, including finance, healthcare, and supply chain management. Figure 8.1 shows the symbol of blockchain [1].

Figure 8.1 The symbol of blockchain [1].

Originally synonymous with cryptocurrencies, blockchain has transcended its financial roots to permeate various industries, promising enhanced security, transparency, and efficiency. Among these sectors, the telecommunications industry stands on the cusp of profound transformation. The telecommunications industry, which is responsible for connecting people and devices globally, can also benefit greatly from blockchain technology. The industry is noted for its complexity, as shown in Figure 8.2 [2].

Figure 8.2 The complexity of telecom infrastructure [2].

Telecom industry today has the most complex operations framework, involving many partners, vendors, customers, distributors, network providers, and VAS providers. The acceptance of blockchain in telecom is a requirement in the digital era. Its transformative potential is reshaping the industry, promising a future of enhanced connectivity and seamless operations. By adopting blockchain technology, companies like China Mobile, Deutsche Telekom, SK Telecom, Telstra, and Vodafone prioritize your security, transparency, and overall satisfaction.

Blockchain is a decentralized peer-to-peer distributed network ledger and the architecture of a new Internet. Blockchain technology has transcended its origins in cryptocurrency to become a pivotal player in the telecommunications sector. It is a game-changer in the telecom industry. Its integration with telecom introduces unparalleled security, transparency, and efficiency, revolutionizing network operations. The integration also represents a significant shift towards more secure, efficient, and transparent operations [3].

In this chapter, we will explore the potential of blockchain in the telecommunications industry. The chapter begins with explaining what blockchain is all about. It describes the role of blockchain in telecommunications. It provides some applications of blockchain in telecommunications. It highlights the benefits and challenges of blockchain in telecommunications. The last section concludes with comments.

8.2 WHAT IS BLOCKCHAIN?

Blockchain, a type of distributed digital ledger technology (DLT), is a relatively new and exciting way of recording transactions in the digital age. It is a decentralized and distributed digital ledger technology that securely records and verifies transactions across multiple computers or nodes in a network. Basically, it is a chain of blocks in which each block contains a list of transactions. The blockchain technology was created as the foundational basis for Bitcoin – a digital currency in which secure peer-to-peer transactions occur over the Internet. It is expected that the spending on blockchain solutions worldwide would grow from 4.5 billion USD (2020) to an estimated value of 19 billion USD by 2024 [4].

Originally developed as the accounting method for the virtual currency Bitcoin, Blockchains are appearing in a variety of commercial applications today. Blockchain technology is a type of distributed digital ledger that uses encryption to make entries permanent and tamper-proof and can be programmed to record financial transactions. It is used for secure transfer of money, assets, and information via a computer network such as the Internet without requiring a third-party intermediary. It is now being adopted across financial and non-financial sectors. As a catalyst for change, the Blockchain technology is going to change the business world and financial matters in major ways.

The first Blockchain was conceived in 2008 by an anonymous person or group known as Satoshi Nakamoto, who published a white paper introducing the concept of a peer-to-peer electronic cash system he called Bitcoin [5,6]. Bitcoin and Ethereum are the first two mainstream blockchains. Other modern blockchains include Namecoin, Peercoin, Ether, and Litecoin. Figure 8.3 shows different components of blockchain [7].

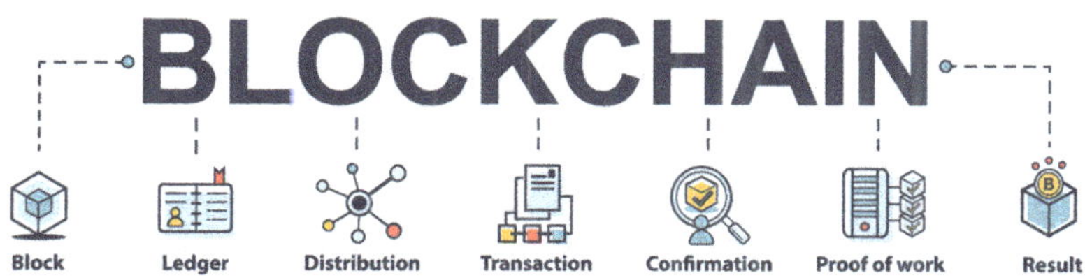

Figure 8.3 Different components of Blockchain [7].

Blockchain combines existing technologies such as distributed digital ledgers, encryption, immutable records management, asset tokenization and decentralized governance to capture and record information that participants in a network need to interact and transact. As illustrated in Figure 8.4, a complete blockchain incorporates all the following five elements [8]:

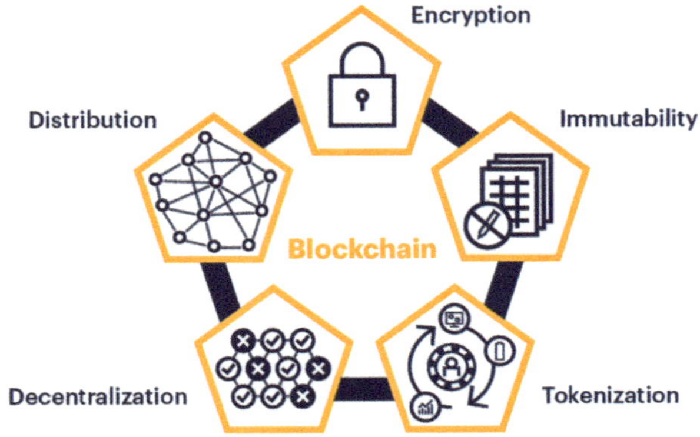

Figure 8.4 Five key elements of Blockchain [8].

- *Distribution:* Digital assets are distributed, not copied or transferred. A protocol establishes a set of rules in the form of distributed mathematical computations that ensures the integrity of the data exchanged among a large number of computing devises without going though a trusted third party. A centralized architecture presents several issues including a single point of failure and problems of scalability.
- *Encryption:* BC uses technologies such as public and private keys to record data securely and semi-anonymously. Completed transactions are cryptographically signed, time-stamped, and sequentially added to the ledger.
- *Immutability:* The Blockchain was designed so these transactions are immutable, i.e. they cannot be deleted. No entity can modify the transaction records. Thus, Blockchains are secure and meddle-free by design. Data can be distributed, but not copied.
- *Tokenization:* Value is exchanged in the form of tokens, which can represent a wide variety of asset types, including monetary assets, units of data or user identities.
- *Decentralization:* No single entity controls a majority of the nodes or dictates the rules. A consensus mechanism verifies and approves transactions, eliminating the need for a central intermediary to govern the network.

Bitcoin and its underlying blockchain technology increasingly impact all facets of society. Bitcoin's status as digital gold is merely the tip of this technology. Figure 8.5 shows Bitcoin [9]. Although blockchain technology will for all time be associated with Bitcoin due to their common genesis, it has broader applications. Cryptocurrency will increasingly become a factor in family law issues as well.

Figure 8.5 Bitcoin [9].

In a nutshell, blockchain technology involves three basic concepts [10]: (1) It is a system for recording a series of data items (such as transactions between parties); (2) It uses cryptography to make it difficult to tamper with past entries; (3) It has an agreed process for storing copies of the ledger and adding new entries (also called a consensus protocol).

8.3 BLOCKCHAIN IN TELECOMMUNICATIONS

Blockchain technology, at its core, is a decentralized digital ledger that records transactions across multiple computers in such a way that the registered transactions cannot be altered retroactively. Blockchain in telecommunications refers to integrating blockchain technology within the telecom industry. It means using unique technology to make telecom networks safer and more efficient. It involves utilizing blockchain's distributed ledger technology to optimize processes, transactions, and security measures within the telecom sector. Figure 8.6 shows blockchain in telecommunications [11].

Figure 8.6 Blockchain in telecommunications [11].

The telecommunications industry plays a pivotal role in global connectivity, and the adoption of blockchain technology has become increasingly relevant due to its potential to address various challenges and provide innovative solutions. Blockchain is being explored by all major industries, including the telecom sector, to determine its benefits and how to apply it to its operations. The integration of blockchain technology in the telecom sector presents numerous opportunities for creating new revenue streams while addressing existing operational challenges. There are many reasons to use blockchain in the telecommunications industry. The telecom sector is suffering from several shortcomings. Blockchain can benefit the telecom industry in overcoming these shortcomings. It has a lot to offer and is ready to deliver results and transform the telecom industry for good. It has the potential to enhance security and create additional sources of income for telecom enterprises. Adopting blockchain is worth the cost.

8.4 APPLICATIONS OF BLOCKCHAIN IN TELECOMMUNICATIONS

Blockchain, which started as the underlying technology for the digital currency Bitcoin, has since expanded to include real estate, finance, education, retail, telecom, and other industries. In the telecom sector, blockchain is already being leveraged to enhance various aspects of operations. Blockchain in telecommunications encompasses various applications that revolutionize the industry, promising enhanced efficiency, security, and transparency. The following applications demonstrate the different ways that telecom industry can invest in the development and implementation of blockchain-based products [12-14]:

- *Decentralized Networks:* Blockchain technology can be used to create a decentralized network for telecommunications providers, which can help to reduce infrastructure costs and improve the speed and reliability of data transfer. The core solution of the telecom problem can be solved by decentralization, which is one of the core features of the blockchain method. Telecom, through blockchain, creates decentralized networks, reducing reliance on central authorities for operations. You benefit from a more resilient, transparent network structure, enhancing reliability and trust in communication.

- *Smart Contracts:* Smart contracts, being self-executing, are programed in a way that they can verify that the terms of the contract have been met and then execute the terms without the need of an intermediary or central authority. Smart contracts, embedded within the blockchain, ensure the integrity of operator transactions while encrypting customer information to prevent tampering. Through blockchain platforms in telecommunications, smart contracts automate and secure agreements. With blockchain technology, telecom providers can create smart contracts that are executed without human intervention, making for a seamless automated experience. You reorganize contract execution, ensuring transparency and reducing transactional disputes. By using blockchain, smart contracts can automate process and guarantee the settlement between the participants, by routing from one operator's blockchain to another operator and increase transparency to the end customer. A telecom provider utilizes smart contracts to automate billing processes for international roaming services.

- *Supply Chain Management:* Telecom operators rely on a complex supply chain to deliver their services, including hardware, software, and third-party services. Blockchain in telecom facilitates and transforms supply chain management. It will enhance transparency and traceability. Moreover, you will track product journeys seamlessly, reducing fraud possibilities and ensuring authentic transactions. Integration of blockchain will simplify supply chain operations, optimizing efficiency and reliability. AT&T is utilizing blockchain to automate and transform their supply chain processes that impact their products – handsets and network equipment.

- *Fraud Prevention:* This is another use case that telcos are depending on blockchain to help cater to. Telecom companies face challenges related to billing accuracy and fraud prevention. They significantly suffer from fraudulent schemes each year. Telecom networks are vulnerable to cyber-attacks and fraud, leading to financial losses and reputational damage. Blockchain can enhance the security of telecom networks by providing a tamper-proof record of all network activities. Using blockchain we can look at significantly decrease the cost of fraud in roaming and in identity management.

- *Billing and Settlement:* Blockchain technology automates critical telecom operations, including billing and roaming. Blockchain modernizes billing processes, ensuring transparent, accurate, and timely payments within the telecom sector. Blockchain technology can be used to create a more efficient and transparent billing and settlement system for telecommunications providers. By using blockchain technology, telecommunications providers can reduce costs and improve the accuracy and speed of billing and settlement processes.

- *Mobile Payment Solutions:* Blockchain technology can be used to create mobile payment solutions that are more secure and efficient than traditional payment methods. By using blockchain technology, telecommunications providers can offer customers a more secure way to make payments and transfer funds. Blockchain for telecommunications revolutionizes mobile payment solutions, enabling secure and efficient transactions. Your mobile transactions become faster, more transparent, and resistant to fraud or manipulation. Telecom enterprises can use blockchain to enable micropayments for music, mobile games, and other services.

- *IoT Connectivity:* Within the next decade, there will be billions of IoT cellular connections. With IoT connections growing exponentially over time, data insecurity will increase simultaneously. Blockchain technology can be used to create a more secure and efficient network for the Internet of things (IoT) devices. By using blockchain technology, telecommunications providers can offer customers a more secure and reliable way to connect their IoT devices to the Internet. Blockchain for telecommunications empowers IoT connectivity in telecom, enhancing the reliability and security of interconnected devices. You experience seamless, trustworthy communication among your IoT devices.

- *Identity Management:* One of the clearest cases of blockchain applications for telecom is identity management. Blockchain can facilitate user-controlled identity management. Users can have control over their personal information, allowing them to share specific data with service providers while maintaining privacy. Currently, when customers register for new accounts, they have to go through a cumbersome "Know Your Customer" (KYC) process which involves handing over or uploading sensitive information. Integrating blockchain with your existing telecom networks can remove this hazard as customer identities are no longer stored with a centralized third-party but in a decentralized, tamper-proof immutable ledger, with no single point of failure. This revolutionary technology gives rise to a blockchain identity management use case (identity-as-a-service).

- *5G Enablement:* 5G is another technology that can benefit from blockchain. The demand for communication services is increasing, and the world is switching to a 5G network, which is ten times faster than 4G. Managing such a complex network will require a greater calculation power and storage capacity. 5G promises prevalent access to various networks, and telecom entrepreneurs will need to handle versatile access nodes. Choosing the fastest access node for every user will soon become a significant challenge for telecom companies. Blockchain has the potential to enable such access selection mechanisms when 5G is developed. It can create highly protected peer-to-peer self-managed mesh networks that use an extensive number of nodes.

• *Operations Advancement:* Currently, the transactions in telecom ledgers need to go through a clearinghouse to be approved. Smart contracts automatically provide the settlement between the participants by routing from one operator's blockchain to another operator. It also increases transparency to the end customer.

• *Content Distribution:* With blockchain for the telecom industry, content distribution becomes more efficient and transparent. You can securely share and access content, ensuring authenticity and copyright protection.

Figure 8.7 shows some of the applications of blockchain in telecommunications [15].

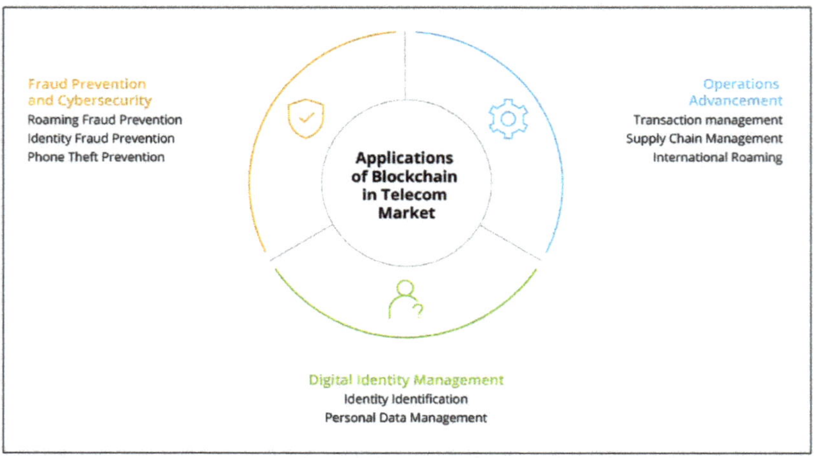

Figure 8.7 Some applications of blockchain in telecommunications [15].

8.5 BENEFITS

The integration of blockchain in telecommunications offers great benefits to telecom. It boosts security by protecting data from online threats, makes transactions faster and cheaper by cutting out middlemen, and ensures everything is transparent and accountable across the network. Other benefits include [13]:

• *Improved Connectivity:* Blockchain can improve connectivity by enabling direct transactions between telecom service providers and their customers. This means that customers can directly pay for the services they use. Blockchain in telecommunications offers enhanced connectivity by creating a decentralized network. You benefit from smoother communication channels, ensuring seamless interactions. With blockchain's distributed ledger system, information transmission becomes more efficient. Whether securing digital identities or managing supply chains, blockchain's integration reshapes telecom into a more efficient, secure, and interconnected ecosystem.

- *Increased Transparency:* There are a lot of trust issues and transparency challenges due to the involvement of multiple entities. Blockchain can increase transparency in telecommunications by providing an immutable and auditable record of all transactions. Blockchain technology in telecom provides heightened transparency within operations. You get a clear view of transaction histories and data exchanges. Through its immutable ledger, every transaction becomes traceable and verifiable. This can help to reduce fraud, improve compliance, and increase trust between telecom service providers and their customers.

- *Greater Security:* In telecom, security is paramount. Cyber security attacks have become a common phenomenon in every industry and the telecommunication industry is not immune to them either. One of the most significant advantages of blockchain in telecommunications is enhanced security. Blockchain technology ensures robust security, safeguarding sensitive data and transactions from cyber threats. Through its decentralized structure, blockchain offers an immutable ledger, minimizing the risk of data breaches. Blockchain helps manage data and transactions better, making systems more reliable and trustworthy. It employs advanced cryptographic techniques to safeguard data integrity. With its decentralized nature, blockchain diminishes vulnerabilities associated with centralized systems.

- *Enhanced Privacy:* Data privacy has been a trending issue for some time now. Awareness among consumers as well as regulators has been increasing about the significance of consumer data. Blockchain technology is safe and robust. It is ideal for storing and processing sensitive information. Blockchain encrypts data and enables users to control their information. It enhances privacy and minimizes the unauthorized access risk.

- *Enhanced Customer Experience:* Through blockchain in telecom, customer experiences are enriched. Transactions become more seamless, empowering users with increased control and transparency over their data and services. Blockchain platforms in telecommunications revolutionize customer experiences by optimizing services. You can enjoy improved service quality and personalized offerings. Through efficient data management, telecom providers modify solutions to your specific needs.

- *Cost Savings:* Blockchain can help telecom service providers to reduce costs by eliminating intermediaries and automating processes. This can help to reduce overheads and improve the efficiency of transactions. Blockchain in telecom enables significant cost savings through restructured processes. By eliminating intermediaries, transactions become more efficient and cost-effective. Smart contracts automate tasks, reducing manual intervention and associated expenses.

- *Improved Interoperability:* Blockchain can improve interoperability between different telecom service providers by creating a decentralized and standardized network. Blockchain technology fosters improved interoperability among telecom systems and devices. It facilitates seamless integration between different networks and platforms. This interoperability enhances communication and data exchange between various stakeholders.

- *Secure Payments:* Blockchain development in telecommunications will ensure secure payments by eliminating intermediaries. You will experience faster, tamper-proof transactions, bolstering trust and reducing fraud risks. Implementing blockchain technology will enhance payment security, streamlining financial processes within the telecom industry.

- *Global Roaming:* At the moment, roaming agreements between telecom operators are complex and require intermediaries to facilitate transactions. With blockchain in telecommunications, global roaming will become easier and transparent. You will experience enhanced connectivity across diverse networks. Moreover, it will reduce complexities and costs associated with roaming. Blockchain will surely facilitate smoother transactions, ensuring better service availability worldwide.

- *Lowering Operational Costs:* Telecommunications operators have long grappled with shrinking margins amidst technological advancements and shifting consumer behaviors. The digitization and virtualization of services have necessitated a reevaluation of operational strategies, with blockchain emerging as a cost-effective solution. The elimination of physical SIM cards and associated distribution costs signifies a symbolic shift towards efficiency and fraud prevention.

- *Quality of Service (QoS):* Blockchain can be used to monitor and verify quality of service metrics, ensuring that customers receive the promised levels of service quality. This is particularly important as 5G technology and the Internet of things (IoT) continue to expand

- *Enhanced Efficiency:* By leveraging smart contracts and automation, blockchain can streamline and optimize various processes involved in international financial transactions. A blockchain-enhanced approach for telecom payment and invoicing offers several advantages over current market norms. It enables interoperability, allowing seamless transactions between different carriers or payment platforms without intermediaries. The use of smart contracts automates processes, increasing efficiency and reducing costs. Blockchain technology improves trust, efficiency, transparency, and security in telecom payment and invoicing systems.

Some of these benefits/advantages are displayed in Figure 8.8 [13].

Figure 8.8 Some of the benefits/advantages of blockchain in telecom [13].

8.6 CHALLENGES

While blockchain brings numerous benefits, challenges exist. Blockchain for telecommunications encounters challenges that necessitate careful consideration and strategic planning. Scalability and interoperability concerns need addressing to ensure seamless integration and widespread adoption across the telecom network. Overcoming integration hurdles, ensuring scalability, complying with regulations, fortifying security, and managing costs are pivotal for successful implementation. Both providers and subscribers struggle with roaming, high fees, risk of fraud, privacy issues, and so much more. Other challenges include the following [13,14]:

- *Integration:* Integrating blockchain for telecommunications poses challenges due to its complex nature. You will need expertise and resources for seamless integration into existing systems. Ensuring compatibility between blockchain and current telecom infrastructure demands careful planning and execution.

- *Scalability:* One of the main challenges of blockchain in telecommunications is its limited scalability. As network demands grow, ensuring blockchain can handle increased transactions and data becomes crucial. One will need scalable solutions that accommodate the expanding network without compromising efficiency.

- *Regulatory Compliance:* Telecommunications companies operate in a heavily regulated environment, and implementing blockchain technology may require them to navigate complex regulatory frameworks. Meeting regulatory requirements poses a challenge for blockchain in telecom. Adhering to evolving legal frameworks while maintaining blockchain's decentralized nature can be intricate and challenging. You must go through complex compliance standards across different regions, ensuring data privacy and security measures align with regulatory mandates.

- *Security:* Although the blockchain platform is renowned for its security, challenges persist within the telecom sector. Safeguarding against sophisticated cyber threats requires constant vigilance and innovation. You will need robust encryption methods and consensus mechanisms to protect sensitive data.

- *Cost:* Implementing blockchain in telecommunications can incur substantial costs. Developing and maintaining blockchain networks demands significant technological, expertise, and infrastructure investments. Upgrading existing systems and training personnel adds to the expenses.

- *Combatting Fraud:* Fraud remains a persistent challenge in the telecommunications sector, particularly concerning roaming services. Through peer-to-peer roaming data exchanges, telecom operators can enhance security measures while minimizing the resources dedicated to fraud detection and prevention.

Some of these challenges are displayed in Figure 8.9 [13].

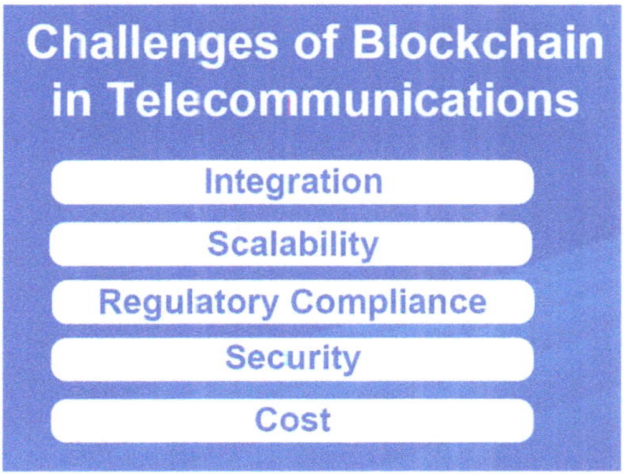

Figure 8.9 Some of the challenges of blockchain in telecom [13].

8.7 CONCLUSION

Although blockchain technology is relatively new to the telecommunications industry, it has the potential of revolutionizing telecommunications in the future. Blockchain is changing how telecom works, offering better security, efficiency, and trust. We are looking at a future where we reduce dependency on traditional centralized software/ operators to provide an authentication mechanism. As the industry undergoes rapid digital transformation, blockchain emerges as a catalyst for innovation, offering solutions to longstanding challenges and unlocking new opportunities for growth and efficiency.

As the need for secure communication grows, picking blockchain for telecom is crucial. It means committing to a safer, faster, and more customer-focused telecom world. The future of blockchain in telecommunications is promising, offering transformative solutions to various industry challenges. Blockchain technology will enable secure and efficient payment systems. More information on the integration of blockchain technology into telecommunications is available from the books in [16-19].

REFERENCE

[1] "Blockchain meets the oil & gas industry," February 2018,

https://executiveacademy.at/en/news/detail/blockchain-meets-the-oil-gas-industry/

[2] S. Chauhan, "Blockchain applications in the telecommunications sector, May 2021,"

https://www.linkedin.com/pulse/blockchain-applications-telecommunications-sector-suman-chauhan/

[3] M. N. O. Sadiku, P. A. Adekunte, and J. O. Sadiku, "Blockchain in telecommunications," *International Journal of Trend in Scientific Research and Development*, vol. 8, no. 6, November-December 2024, pp. 378-388.

[4] C. M. M. Kotteti and M. N. O. Sadiku, "Blockchain technology," *International Journal of Trend in Research and Development*, vol 10, no. 3, May-June 2023, pp. 274-276.

[5] "Blockchain," *Wikipedia*, the free encyclopedia

https://en.wikipedia.org/wiki/Blockchain

[6] S. Nakamoto, "Bitcoin: A peer-to-peer electronic cash system,"

https://bitcoin.org/bitcoin.pdf

[7] "The beginning of a new era in technology: Blockchain traceability,"

https://www.visiott.com/blog/blockchain-traceability/#:~:text=The%20Beginning%20of%20a%20New,money%20without%20a%20central%20bank.

[8] "The CIO's guide to blockchain,"

https://www.gartner.com/smarterwithgartner/the-cios-guide-to-blockchain#:~:text=True%20blockchain%20has%20five%20elements,%2C%20immutability%2C%20tokenization%20and%20decentralization.

[9] “Blockchain and space exploration: Is decentralized data the future of space missions?” October 2024,

https://medium.com/coinmonks/is-decentralized-data-the-future-of-space-missions-646173d1aeec

[10] D. Michels, “Technology blockchain and telecoms,”

chrome-extension://efaidnbmnnnibpcajpcglclefindmkaj/https://www.iicom.org/wp-content/uploads/22-26-blockchain.pdf

[11] K. Leslove, “Do telecom firms need to focus on blockchain integration?” December 2022,

https://media.hashcashconsultants.com/do-telecom-firms-need-to-focus-on-blockchain-integration/

[12] “Top 5 blockchain use cases in telecom,”

https://www.protokol.com/insights/top-5-blockchain-use-cases-in-telecom/

[13] “Blockchain for telecommunications: Innovative industry solution,” January 2024,

https://webisoft.com/articles/blockchain-for-telecommunications/

[14] H. White, “5 Ways in which blockchain will revolutionise telecom industry,” March 2024,

https://www.mobiliseglobal.com/five-ways-blockchain-will-change-telecoms-industry/

[15] “Blockchain Use Cases in the Telecom Industry,” January 2023,

https://www.infopulse.com/blog/blockchain-use-cases-in-the-telecom-industry

[16] M. N. O. Sadiku, *Blockchain Technology and Its Applications*. Moldova, Europe: Lambert Academic Publishing, 2023.

[17] M. Pustišek, N. Živić, and A. Kos, *Blockchain: Technology and Applications for Industry 4.0, Smart Energy, and Smart Cities (De Gruyter STEM)*. De Gruyter, 2021.

[18] M. H. Rehmani, *Blockchain Systems and Communication Networks: From Concepts to Implementation (Textbooks in Telecommunication Engineering)*. Springer, 2021.

[19] S. S. Biswas, *Blockchain in Telecomn An Insight into the Potential Benefits of Combining Blockchain, 5G, IoT, Cloud Computing, and AI/ML in the Telecom Business (English Edition)*. Bpb Publications, 2021.

CHAPTER 9
5G NETWORK IN TELECOMMUNICATIONS

"5G will have an impact similar to the introduction of electricity or the car, affecting entire economies and benefiting entire societies."

—Steve Mollenkopf

9.1 INTRODUCTION

The introduction of 5G technology is set to revolutionize the way we communicate, work, and live. 5G stands for the fifth generation of cellular networking, surpassing 4G LTE. 5G technology is developing rapidly to enable the next generation of wireless communications and pave the way for more than just super-fast data transfer. The 5G wireless technology will usher in an era of gigabit speeds for everyone, reducing latency for customers and organizations. The advent of 5G technology will not only transform our devices but will also significantly change the way businesses operate, communicate, connect, and share data. 5G is expected to be a game-changer in the business world [1]. Figure 9.1 shows the symbol of 5G [2].

Figure 9.1 The symbol of 5G [2].

5G is the newest generation in wireless networking, with greater capacity and higher speed. It will integrate with infrastructure, buildings, and appliances, creating an Internet of things (IoT). It supports interactive multimedia, voice, video, Internet, and other broadband services. To support increased throughput requirements of 5G, a new spectrum has been assigned to 5G in mmWave bands. 5G will use multiple inputs multiple outputs (MIMO) to significantly increase network capacity [3-5].

The telecom industry continues to be a beacon of digital transformation, integrating innovative technologies to enhance connectivity and communication. The telecom infrastructure is known for its complexity, as typically shown in Figure 9.2 [6].

Figure 9.2 The telecom infrastructure is known for its complexity [6].

5G connectivity promises to break traditional data delivery paradigms by providing network connectivity almost everywhere. The capabilities of 5G have opened opportunities for the telco industry to serve customers closer to the edge of the network with more solutions. Americans all over the nation are using 5G to create the future and enhance America's economic competitiveness. 5G is connecting students to learning opportunities, making farms more sustainable, fighting wildfires, making manufacturing smart, and driverless cars a reality.

In today's fast-paced digital world, mobile communication has become an integral part of our daily lives. 5G represents a significant leap forward in the evolution of mobile communications, unlocking new possibilities for connectivity, productivity, and technological innovation worldwide. It is the latest innovation in wireless technology that promises to deliver faster speeds, lower latency, and greater reliability than its predecessors. Central to its appeal is its capacity to handle vast volumes of data generated by increasingly complex devices connected to its networks. It is a game-changer across various industries including telecommunications. Although this upgrade will be expensive, it is essential to beat the competition [7].

This chapter explores the applications of 5G networks in telecommunications. It begins with providing some background information about 5G network. It discusses the role of 5G network in telecommunications. It covers some applications of 5G network in telecommunications. It highlights the benefits and challenges of 5G network in telecommunications. The last section concludes with comments.

9.2 OVERVIEW OF 5G NETWORK

5G is the fifth-generation wireless cellular technology that will provide faster and more reliable communication with low latency. Compared to its predecessor, it is estimated that the 5G mobile network allows 1,000 times more data transmission compared to 4G.

Like its predecessors—3G, 4G, and 4G —5G utilizes radio waves to transmit data.

Evolution from 1G to 5G is depicted in Figure 9.3 [8], while the relationship between 3G, 4G, and 5G is portrayed in Figure 9.4 [9].

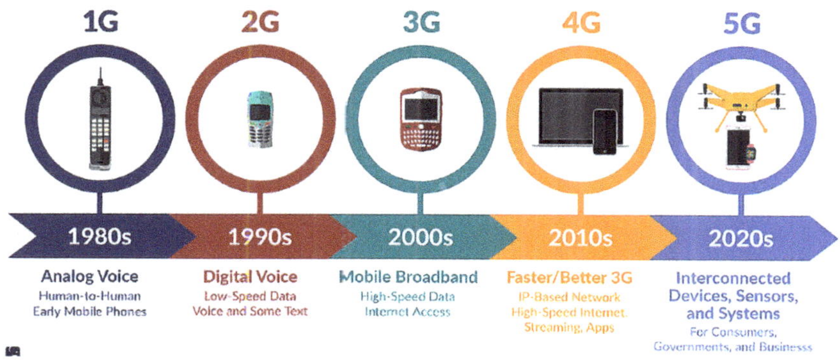

Figure 9.3 Evolution from 1G to 5G [8].

Figure 9.4 Relationship between 3G, 4G, and 5G [9].

Like any other cellular network, 5G networks send data through radio waves and operate on a cellular infrastructure, where geographic regions are partitioned into cells, each supported by an antenna and a base station. Each cell is connected to a network backbone through a wired or wireless connection. 5G may transmit data over the unlicensed frequencies currently used for Wi-Fi. It promises a smarter, faster, and more efficient network. The goal of 5G is to have far higher speeds available, at higher capacity per sector, and at far lower latency than 4G. To increase network efficiency, the cell is subdivided into micro and pico cells [10]. 5G will be a new mobile revolution as it is expected to provide gigabit-per-second data rates anytime, anywhere. 5G uses towers, as typically shown in Figure 9.5 [11].

Figure 9.5 A typical 5G towers [11].

5G towers are telecommunications sites capable of transmitting 5G signals for wide-area coverage. 5G cell towers use a combination of low, mid, and high-frequency bands for various connectivity use cases. Towers themselves are not 5G; it is the equipment on the tower that makes it 5G. Figure 9.6 shows how 5G works [12].

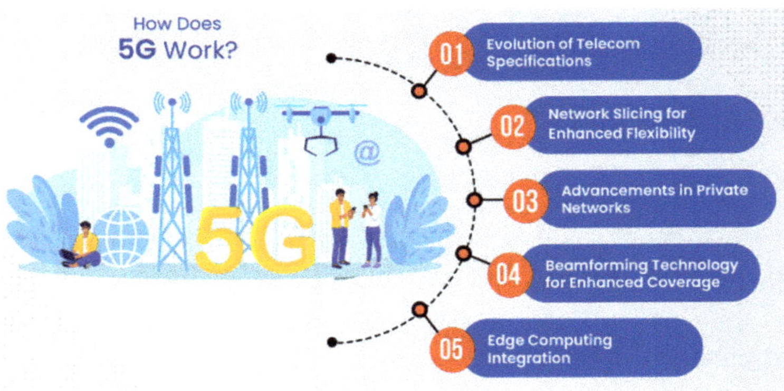

Figure 9.6 How 5G works [12].

In a 5G wireless network, every mobile phone will have an IPv6 address depending on the location and network being used. 5G utilizes the user-centric network concept World Wide Wireless Web (WWWW) instead of operator-centric as in 3G or service-centric as in 4G [13]. WWWW will be capable of supporting applications and services and interconnecting the whole world. 5G includes the latest technologies, such as cognitive radio, the Internet of things, nanotechnology, and cloud computing.

The key features of 5G include high throughput, improved spectrum efficiency, reduced latency, better mobility support, and high connection density. 5G technology has the following advanced features [14]:

- Architecture will be device-centric, distributed, programmable, and cloud-based
- High data rates
- One to 10 Gbps connections to endpoints
- One millisecond end-to-end round trip delay
- Low battery consumption
- Better connectivity irrespective of location
- Larger number of supporting devices
- Lower cost of infrastructure development

Some of these features are illustrated in Figure 9.7 [15].

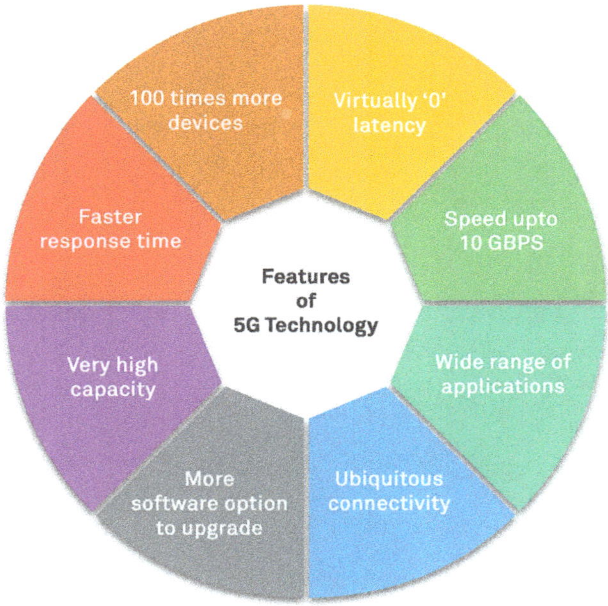

Figure 9.7 Some of the features of 5G [15].

The development of 5G will not be from scratch but will gradually build on 4G LTE. Major technologies enabling 5G include:

- *D2D Communication:* Direct connectively is achieved through device-to-device (D2D) technology. 5G cellular network will implement D2D mm wave communication technology to provide high-speed data rate, improve coverage, and offer peer-to-peer services. Much research has been invested in characterizing D2D connections as part of LTE [16].
- *M2M Communication:* While D3D communication targets mobile radios, machine-to-machine (M2M) expands the scope and facilitates ubiquitous connectivity among mobile devices. It is estimated that there will be over 100 billion connected devices using M2M communications in the 5G backbone [17].
- *MIMO:* Multiple-input-multiple-output (MIMO) technology plays a crucial role in 4G and is expected to play an important function in 5G. Massive MIMO extracts the benefits of

MIMO on a large scale by increasing the throughput and spectrum efficiency.

Other enabling technologies of 5G include mmWave communication, ultra-dense network (UDN), all-spectrum access (ASA), OFDM (orthogonal frequency division multiplexing), and the Internet of things. Industries that use 5G technology are shown in Figure 9.8 [8].

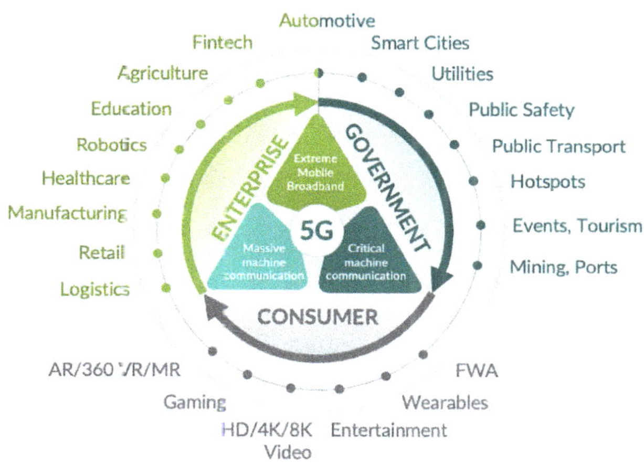

Figure 9.8 Industries that use 5G technology [8].

This section would be incomplete without mentioning the successors of 5G [18]:

• *6G Network:* Fifth-generation cellular technology is replaced by sixth-generation wireless or 6G. The bandwidth and latency of 6G networks will be significantly higher than those of 5G networks due to their ability to operate at higher frequencies. The main purpose of 6G internet is to provide communications with one-microsecond latency. 6G will employ satellites to connect the current 5G networks.

• *7G Network:* Globally, the 7G Network provides a faster means of communication. The advanced cellular technology that will be the successor for 5G and 6G. A 7G network is the quickest way to make a call, whether it is local or international. Voice over

Internet Protocol (VoIP), or 7G, requires access to all local and international telecommunications. 7G will be able to satisfy the requirements of extremely high bandwidth, almost zero latency, and universal integration. Although 7G will not be generally available until 2030, a handful of countries are currently using it. These include Norway, Netherlands, South Korea, and Hungary. They are the nations in the world to provide the fastest Internet speeds.

9.3 5G NETWORK IN TELECOMMUNICATIONS

5G technology is one of the main forces behind the ongoing change and evolution of the telecom sector. It is positioned to become the standard for mobile communications technology, ushering in a new era of connectivity, speed, and possibility. 5G is poised to connect businesses, people, and processes at enormous scale. It will enable new and creative use cases that are not feasible with today's wireless networks.

5G signifies a significant advancement for the telecom sector. It provides greater reliability, reduced latency, and faster speeds, all of which are necessary for the upcoming creation of new services and apps. The telecom sector needs to engage in 5G if it wants to remain competitive. Figure 9.9 shows a telecom field engineer [19].

Figure 9.9 A telecom field engineer [19].

9.4 APPLICATION OF 5G NETWORK IN TELECOMMUNICATIONS

5G is the fifth generation technology standard for broadband in the telecommunications industry and the successor to the 4G networks. 5G networks support applications requiring low latency, like industrial automation, augmented reality, and autonomous vehicles, with unprecedented efficiency and responsiveness. 5G in telecommunications can be applied in the following ways [20]:

- *Smart Roads:* Telecommunications companies can collaborate with municipalities to implement the smart roads project. This is a technological plan that enhances road safety by using special sensors placed on crosswalks to inform passing cars that a pedestrian is approaching. This is made possible due to 5G technology. By reducing latency times to a minimum, information can be immediately exchanged between multiple devices simultaneously.

- *Smart Cars:* The advent of 5G technology has revolutionized the field of autonomous vehicles, encompassing everything from self-driving cars to drones. With the unparalleled connection speeds of 5G, transport systems for various modes of transportation, including cars and trains, have experienced a substantial acceleration in performance. For the automotive industry to implement smart cars, it needs to collaborate with telecommunications companies. Autonomous driving vehicles need to rely on the 5G network since it is the most efficient technology for transferring large amounts of data quickly, constantly, and in a stable way. These vehicles will be able to communicate with each other to avoid accidents and ease traffic, thanks to the quick and reliable data 5G can send.

- *Smart Cities:* City life will become more efficient as 5G helps manage traffic flows, public transport, and energy use, making cities cleaner and less congested.

- *Home Automation:* 5G in the telecommunications sector is key to expanding home automation. This includes the installation and usage of smart home devices such as refrigerators, blinds, doors, light bulbs, heaters, ovens, and other home appliances that can communicate with each other and mobile devices for control and monitoring. Many countries in Europe have implemented a home automation tax benefit that is available, which provides tax deductions for smart home devices.

9.5 BENEFITS

5G will usher in a transformative era, particularly unlocking new possibilities in urban settings where over two-thirds of the global population is estimated to reside by 2050. There are many ways that telecommunications companies can reap the benefits from 5G's low latency and higher transmission speeds. As the benefits of 5G have become more accessible, businesses are planning to use 5G in their operations, particularly in rural and field connectivity. Other benefits include the following [2,21]:

- *Increased Speed:* One of the most exciting benefits of 5G is its faster upload and download speeds. We can expect to see 5G reach up to 10 Gbps, up to 100 times faster than 4G. The increased speed of 5G mobile networks introduces exciting possibilities for both consumers and businesses alike. This will enable activities like streaming video, downloading audio, browsing the web, and using social media.

- *Enhanced Capacity:* One of the significant benefits of 5G is its ability to support a massive number of connected devices simultaneously. This capability is crucial for the growth of the Internet of things (IoT) and the deployment of smart devices. With the ability to provide up to 1,000 times more capacity than 4G, 5G will take IoT development to the next level and pave the way for smart cities and homes. 5G's increased capacity has huge potential for industry.

- *Lower Latency:* A standout feature of 5G is its substantially reduced latency, which measures the time required for data to travel from its original to its target location. Each generation of cellular communications technology has strived to reduce latency, and 5G is no different. 5G is expected to transmit data in less than five milliseconds — virtually no time at all. This dramatic latency reduction enhances real-time responsiveness, facilitating tasks such as file downloads and cloud-based operations with remarkable speed and efficiency. 5G's lower latency will also enable users to control devices remotely in real-time. Reduced latency will transform how we create, store, use, and share data in the business sector. 5G's lower latency will also enable users to control devices remotely in real-time, offering enormous benefits for businesses with multiple or remote locations.

- *Increased Bandwidth:* By increasing speeds and network capacity, 5G will also open up the doors to greater bandwidth than with 4G. This enables them to optimize network traffic better. For businesses, 5G's increased bandwidth will help them gain access to big data.

- *Improved Coverage:* One of the subtler benefits of 5G is its frequency, which can provide speeds much faster than 4G and even pass through buildings to provide greater coverage indoors and outdoors. Although many telecommunications providers are rapidly expanding coverage across the country, 5G is not yet available everywhere. By combining the latest 5G technology with satellite Internet, you can rest easy knowing your systems will stay online 24/7, even when cellular connections are not possible.

- *Enhanced Communication:* For businesses that already use IoT, AI, and machine learning, the introduction of 5G will be a massive boost to their efficiency. Cloud computing is a critical part of any modern business's IT infrastructure, allowing for the simple storage and transfer of large amounts of data between devices. With the addition of 5G, the cloud will become even more integral to how businesses rely on their networks.

- *Improved Connectivity:* 5G is the new standard in mobile wireless networking that brings improved Internet connectivity in a digital-first world. 5G technology will allow for more devices to connect to the Internet simultaneously, with improved network coverage and capacity. This will enable the Internet of things (IoT) to become more widespread, with a greater number of connected devices and sensors. There is no need for cables or fiber optic lines when using 5G to provide high-speed Internet connectivity to homes and businesses. 5G network seamlessly connects everyone and everything worldwide.

- *Increased Revenue:* The increased speed, connectivity, and capacity of 5G networks will create new revenue opportunities for telecom operators, such as offering premium services to businesses or charging for access to high-speed networks.

- *Customer Satisfaction:* With market saturation and a high average churn rate, telcos have been under increased pressure over the past decade to reinvent themselves as digital service providers. Customers are looking for a unified digital experience from a single trusted provider who can offer services and products in a one-stop shop. This is opening an opportunity for providers to complement core services with digital services. For example, a T-Mobile customer can get mobile service, along with subscriptions to streaming services for shows and music, on one bill.

- *Enabler for Emerging Technologies:* The demand for faster speeds to support emerging technologies such as artificial intelligence (AI), the Internet of things (IoT), and machine learning (ML) has escalated beyond the capabilities of 4G network. Those technologies will be fully developed through the realization of 5G. This is because these technologies require much higher processing powers for the constant streaming of massive volumes of data. 5G is expected to enable new applications such as augmented reality, virtual reality, and smart cities, and transform industries such as healthcare, transportation, and manufacturing.

- *Investing In the Future:* With 5G, telcos can take advantage of three distinct markets faster — cybersecurity, cloud services, and the Internet of things (IoT). Telcos aiming to thrive in these markets have used a two-pronged approach of freeing up cash, then re-investing those funds into low-latency infrastructure (fiber optic network and 5G) and cloud services companies.

9.6 CHALLENGES

The implementation of 5G is not without its challenges, including the need for significant investment in infrastructure, cost to consumers, spectrum availability, security concerns, interoperability, infrastructure investment, spectrum availability, network security, and regulatory considerations. Besides, the challenges that hinder 5G infrastructure development include intense competition for higher spectrum bands, the need for numerous antennas and base stations due to interference issues, the high costs of upgrading infrastructure, navigating complex government regulations for cybersecurity and spectrum availability, and ensuring ultra-secure cloud-based and data virtualization services to defend against cybersecurity threats associated with increased connectivity [22]. Other challenges include the following [23]:

- *Security:* 5G network is managed by software, which is easily susceptible to vulnerabilities. And this creates a path that is accessible by hackers to breach privacy, systems, data and much more. Due to the growing number of connected devices and the use of new technologies, 5G networks are susceptible to security risks like cyberattacks and hacking. New security protocols and steps are needed to ensure the security of 5G networks.

- *Regulatory Compliance:* Telecommunications firms may find it difficult and time-consuming to comply with the stringent regulatory compliance standards that apply to the rollout of 5G. This covers observing data privacy laws, network security requirements, and other rules pertaining to the installation and use of 5G networks.

- *Cost:* Consumers may pay more for 5G technology than for earlier wireless technology generations. This may limit the use of 5G in some markets and make it impossible for some users to finance the new technology.

- *Confusion:* Although the "race to 5G" has been widely publicized, the fundamental issues and realities underpinning the transition to 5G technology are still widely misunderstood. Many people find the transitions by the broadband cellular networks from 3G to 4G and now to 5G confusing and do not understand the reasons behind these changes. This issue of EUC will address some of this confusion and discuss a number of ways that new generations cellular broadband networks and devices are important to virtually every industrial and commercial business.

- *Deployment:* Deploying 5G infrastructure is challenging because high-frequency waves, which can carry lots of data, are easily blocked by buildings, trees, and weather. To overcome this, a dense network of small antennas needs to be installed on streetlights, buildings, and other structures, rather than relying on traditional cell towers. In the hardware segment, radio access network (RAN) dominated the 5G infrastructure market with a share of 50%. The trend of deploying virtual and centralized RAN is rapidly increasing among network service providers to reduce the overall infrastructure costs and network complexities.

- *Decline:* The telecoms industry is arguing that a "perpetual decline in terms of profitability" and this is affecting the sector's ability to deliver high-speed connectivity - an issue that big tech can help ameliorate by contributing to the costs of network infrastructure. Recent figures illustrate how the industry has been displaying symptoms of a longer-term decline.

9.7 CONCLUSION

5G represents a monumental leap forward in wireless technology, offering unparalleled capabilities to meet the escalating demands of the digital era. It is positioned to become the standard for mobile communications technology, ushering in a new era of connectivity, speed, and bandwidth. Businesses and consumers may finally have it all: high capacity, ultra-reliability, low latency, reduced energy usage, and massive connectivity.

Many nations are launching 5G networks, making services increasingly available to consumers. The integration of 5G with other emerging technologies such as artificial intelligence, machine learning, and the Internet of things (IoT) will unlock new opportunities for innovation and disruption. We can expect to see greater advances in areas such as virtual and augmented reality, smart cities, and autonomous vehicles.

The future of 5G in the telecommunications industry is highly promising and holds immense potential to revolutionize multiple industries including healthcare, transportation, manufacturing, entertainment, and telecommunications. More information on the implementation of 5G networks in the telecommunications industry is available from the books in [24-27].

REFERENCE

[1] M. N. O. Sadiku, U. C. Chukwu, and J. O. Sadiku, "5G for business," *International Journal of Human Computing Studies*, vol. 5, no. 5, May 2023, pp. 61-68.

[2] J. Warrington, "How 5G is reshaping the telco industry and its impact on customer retention," June 2023,

https://www.channelfutures.com/mobility-wireless/how-5g-is-reshaping-the-telco-industry-and-its-impact-on-customer-retention

[3] "5G," *Wikipedia*, the free encyclopedia

https://en.wikipedia.org/wiki/5G

[4] M. N. O. Sadiku, K. G. Eze, and S. M. Musa, "5G wireless technology: A primer," *International Journal of Scientific Engineering and Technology*, vol. 7, no. 7, July 2018, pp. 62-64.

[5] M. N. O. Sadiku, C. M. Kotteti, and S. M. Musa, "5G technology: A primer," *International Journal of Trend in Research and Development*, vol. 6, no. 1, Jan.-Feb. 2019, pp. 382-384.

[6] D. Fellers, "Digital solutions economy in the telecom industry," January 2023,

https://ignitepossible.bramasol.com/blog/digital-solutions-economy-in-the-telecom-industry

[7] M. N. O. Sadiku, P. A. Adekunte, and J. O. Sadiku, "5G network in telecommunications," *International Journal of Trend in Scientific Research and Development*, vol. 8, no. 6, November-December 2024, pp. 370-378.

[8] "[Burning issue] 5G technology,"

https://www.civilsdaily.com/burning-issue-5g-technology-2/

[9] "5G technology and networks (speed, use cases, rollout)," September 2024,

https://www.thalesgroup.com/en/markets/digital-identity-and-security/mobile/inspired/5G#:~:text=5G%20connectivity%20promises%20to%20break,in%20delivery%2C%20and%20complete%20interoperability.

[10] R. S. Sapakal and S. S. Kadam, "5G mobile technology," *International Journal of Advanced Research in Computer Engineering & Technology*, vol. 2, no. 2, February 2013, pp. 568-571.

[11] K. Kimachia, "Top 8 5G issues and disadvantages to know before switching," June 2023,

https://www.enterprisenetworkingplanet.com/management/top-issues-facing-5g-networks/

[12] "5G use cases paving the way for technological advancements,"

https://www.veritis.com/blog/5g-use-cases-paving-way-for-technological-advancements/

[13] "How 5G technology works,"

https://www.rfpage.com/how-5g-technology-works/

[14] D. H. Latha et al., "A study on 5th generation mobile technology – Future network service," *International Journal of Computer Science and Information Technologies*, vol. 5, no. 6, 2014, pp. 8309-8313.

[15] "Guest blog: Business @ speed of thought: The power of 5G, Edge, and IoT," December 2020,

https://www.techuk.org/resource/guest-blog-business-speed-of-thought-the-power-of-5g-edge-and-iot.html

[16] S. Andreev, "Delivering uniform connectively and service experience to converged 5G wireless networks," *Proceedings of the IEEE World Forum on Internet of Things*, 2014, pp. 323-324.

[17] M. H. Alsharif et al., "How to make key wireless technologies environmental friendly: a review," *Transactions of Emerging Telecommunications Technologies*, vol. 29, 2018.

[18] A. Kayyali, "Countries using 7G network technology today," October 2022,

https://insidetelecom.com/countries-using-7g-network-technology-today/

[19] "The future of 5G & AI in telecommunications," May 2024,

https://www.ni.com/en/solutions/5g-6g/future-5g-ai-telecommunications.html?srsltid=AfmBOoq9hI_i9xH_GNpoV5JqzzG0NXU2lXcenWke98TfioJ1wFChNKMF

[20] "Ways that telecommunications companies can benefit from 5G,"

https://blinqnetworks.com/ways-that-telecommunications-companies-can-benefit-from-5g/

[21] Human Marketing, "How will 5G improve the telecommunications industry?" November 2021,

https://www.ipinternational.net/how-will-5g-improve-the-telecommunications-industry/

[22] "Understanding 5G infrastructure - New trends in technology,"

https://www.c3controls.com/white-paper/understanding-5g-infrastructure-new-trends-in-technology/

[23] R. Niranjana, "5G's impact on telecom industry: Advantages, challenges, applications, & advancements," May 2023,

https://www.logic-fruit.com/blog/wireless-communication/5g-impact-on-telecom/?srsltid=AfmBOopqSpDij2NvQZw9fezKTeZjZWrlmQP93pfJTYxWzAxa8E88u3ED

[24] W. Jiang and B. Han, *Cellular Communication Networks and Standards: The Evolution from 1G to 6G*. Springer, 2024.

[25] A. F. Molisch, *Wireless Communications: From Fundamentals to Beyond 5G*. Wiley-IEEE Press; 3rd edition, 2022.

[26] E. Markakis et al. (eds.), *Cloud and Fog Computing in 5G Mobile Networks: Emerging advances and applications*

(Telecommunications). The Institution of Engineering and Technology, 2017.

[27] W. Chen et al., *Fundamentals of 5G Communications: Connectivity for Enhanced Mobile Broadband and Beyond.* McGraw Hill LLC, 2021.

CHAPTER 10
3D PRINTING IN TELECOMMUNICATIONS

"3D printing will change the world by allowing for mass customization and on-demand production."

—Avi Reichenhall

10.1 INTRODUCTION

The advent of 3D printing technology has emerged as a game-changer in numerous industries, completely transforming conventional methods of production. This technology is growing and can be used for small productions of these computer-made designs in many industries, including in the telecom industry. With the ability to create complex parts and products quickly and efficiently, 3D printing is transforming the way that telecom companies operate [1]. However, it has been predominantly employed in rapid-prototyping and tooling rather than actual production. It is now capable of manufacturing complex and intricate designs, allowing for customization and personalization in ways previously unimaginable. The level of adoption of 3D printing varies greatly across different sectors. The adoption rates of the technology across eight key industries are depicted in Figure 10.1 [2].

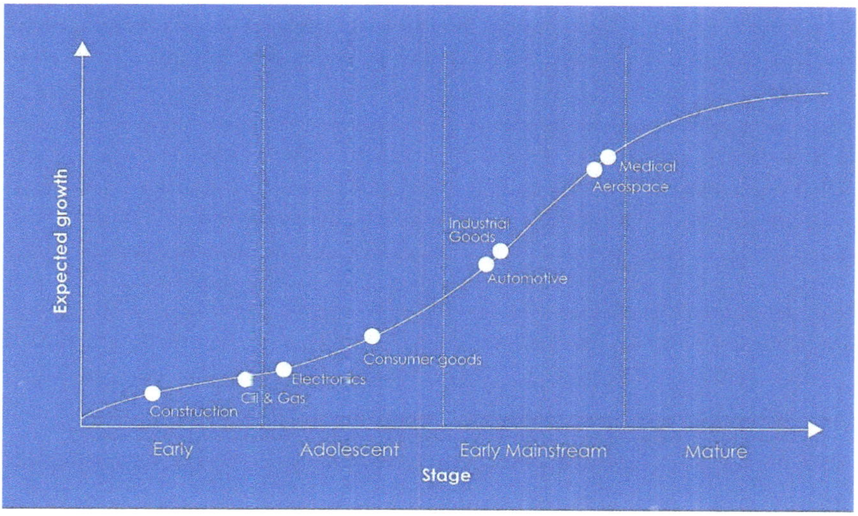

Figure 10.1 The adoption rates of the technology across eight key industries [2].

The telecommunication activity is everywhere, and we rely on it evermore throughout our daily lives. It connects people around the globe. Telecommunication technologies have become an inevitable part of our day-to-day lives from mobile connectivity to home connectivity, office conferencing to remote surgeries. Telecommunication infrastructure is noted for its complexity, as typically shown in Figure 10.2 [3].

Figure 10.2 Telecommunication infrastructure is noted for its complexity [3].

In spite of the ever-deepening reach of communications technologies, users still hunger for greater speeds, more reliable connectivity, and high-performing devices to enhance experiences. There is always room for improvements.

3D printing is a digital manufacturing technology that can create physical objects from a digital file like CAD. This innovative technology allows objects to be created layer by layer, revolutionizing the traditional manufacturing process. It has emerged as a disruptive force in various sectors, including manufacturing, healthcare, and aerospace. It has the potential for revolutionizing the telecommunications industry [4].

The chapter examines the applications of 3D printing in the telecommunication industry. It begins with explaining what 3D printing is all about. It presents different types of 3D printing. It covers 3D printing in telecommunications and provides some of its applications. It highlights some benefits and challenges of 3D printing in telecommunications. The last section concludes with comments.

10.2 WHAT IS 3D PRINTING?

3D printing (also known as additive manufacturing (AM) or rapid prototyping (RP)) was invented in the early 1980s by Charles Hull, who is regarded as the father of 3D printing. Since then it has been used in manufacturing, automotive, electronics, aviation, aerospace, aeronautics, engineering, architecture, pharmaceutics, consumer products, education, entertainment, medicine, space missions, military, chemical industry, maritime industry, food industry and jewelry industry. All the parts created using a 3D printer need to be designed using some kind of CAD software [5,6].

The process of 3D printing has three basic components: computer assisted (i.e. digital) design, machine equipment, and an added material. Different materials can be used to build 3D-printed objects, from plastic, metal and rubber to human cells. A 3D printer works in similar ways as a regular printer. 3D Printing essentially describes a assortment of technologies that digitally formulate three dimensional objects on a layer-by-layer basis. It has been adopted by students, entrepreneurs, hobbyists, and various industries. As shown in Figure 10.3, 3D printing involves three steps [7]. A typical 3D printer is shown in Figure 10.4 [8]. The business uses of 3D printing are growing year by year.

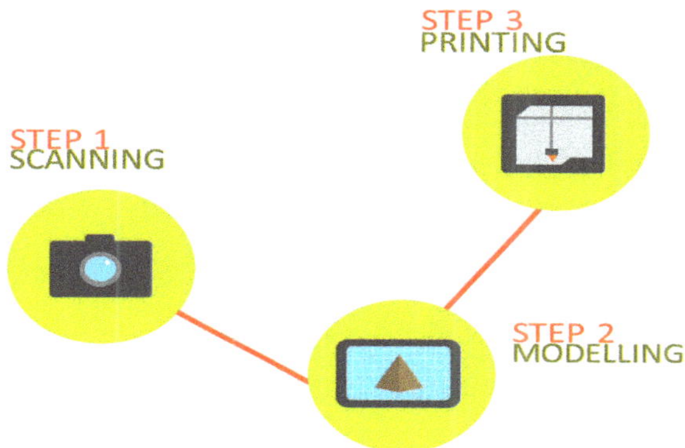

Figure 10.3 3D printing involves three steps [7].

Figure 10.4 A typical 3D printer [8].

In essence, 3D printing is a manufacturing process in which material is laid down, layer by layer, to form a three-dimensional object. 3D printing can create physical objects from a geometrical representation by successive addition of material. 3D printing is an umbrella concept for a set of processes and technologies that offer a wide range of the production of parts and products in different materials. Varieties of 3D printing technologies have been developed with different functions. One thing common in all these processes is the manner in which production is carried out – layer by layer in an additive process. Not all 3D printers use the same technology.

Depending on the technology, 3D printers can use a variety of materials, including but not limited to metals (stainless steel, solder, aluminum, and titanium among them); plastics and polymers (including composites that combine plastics with metals, wood, and other materials); ceramics; plaster; glass; and even foodstuffs like cheese, icing, and chocolate. Even composites can be used. The choice of material greatly depends on the product's function and desired properties [9,10].

10.3 TYPES OF 3D PRINTING

There are generally three types of additive manufacturing: selective binding, selective solidification, and selective deposition. Typically, people refer to these technologies as Selective Laser Sintering (SLS), Stereolithographic (SLA), and Fused Deposition Modeling (FDM), which are discussed as follows [11,12]:

- *Stereolithography (SLA):* This was the world's first 3D printing technology, invented in the 1980s. It is an additive manufacturing process which employs a vat of liquid ultraviolet curable photopolymer "resin" and an ultraviolet laser to build parts' layers one at a time. For each layer, the laser beam traces a cross-section of the part pattern on the surface of the liquid resin. SLA parts have the highest resolution and accuracy and the smoothest surface finish of all plastic 3D printing technologies. Although stereolithography can produce a wide variety of shapes, it has often been expensive.

- *Selective laser sintering (SLS):* This is an additive manufacturing technique that uses a high power laser (for example, a carbon dioxide laser) to fuse small particles of plastic, metal (ceramic, or glass powders into a mass that has a three-dimensional shape. The SLS machine preheats the bulk powder material in the powder bed somewhat below its melting point. SLS is trusted by engineers and manufacturers across different industries for its ability to produce strong, functional parts. Low cost per part, high productivity, and established materials make the technology ideal for a custom manufacturing. The material selection for SLS is limited compared to FDM and SLA.

- *Fused Deposition Modeling (FDM):* This is the most widely used form of 3D printing at the consumer level, fueled by the interest of hobbyists of 3D printers. This technique is suited for basic proof-of-concept models, as well as quick and low-cost prototyping of simple parts. FDM is regarded as a very clean technology, usually simple and office-friendly. It uses a continuous filament of a thermoplastic material. The technology can produce complex geometries and cavities that would otherwise be quite problematic. Since 2004, FDM technology has been used in a particular sector to produce load-bearing scaffold. Home printer based on FDM typically work with plastic filament.

Which technology makes the most sense for you to use depends on your budget, the model's complexity, and the finest detail that is necessary. In the telecom 3D printing process, we use FDM and SLA.

10.4 3D PRINTING IN TELECOMMUNICATIONS

The telecom industry has been at the forefront of innovation and technological advancement for decades. From the invention of the telephone to the rise of the Internet and wireless communication, the industry has continuously pushed the boundaries of what is possible. 3D printing is being used in the telecom industry for a variety of applications, from prototyping to producing end-use parts and creating customized products. Telecom-driven 3D printing is a novel concept that combines two cutting-edge technologies to unleash a myriad of possibilities. While 3D printing allows the creation of three-dimensional objects by adding layer upon layer of material, telecommunications provide the high-speed connectivity required for seamless transmission of digital information. The synergy between telecommunications and 3D printing is paving the way for innovative breakthroughs in various industries. By embracing this convergence of telecommunications and 3D printing, businesses can unlock new opportunities, streamline their processes, and gain a competitive edge. As the market continues to grow, telecommunication sector must embrace this transformative technology to stay competitive and innovative [13].

10.5 APPLICATIONS OF 3D PRINTING IN TELECOMMUNICATIONS

Telecommunications are the unsung heroes of the 3D printing revolution. They keep everything connected and running smoothly behind the scenes. The 3D printed design could be a game changer for the telecommunications industry. 3D printing has been adopted by telecom companies for a variety of applications, from prototyping to producing end-use parts. 3D printing is applied in telecommunications in the following ways [13,14]:

- *Rapid Prototyping:* One of the most significant ways in which 3D printing is used in the telecom industry is for prototyping. The beauty of 3D printing lies in its transformative capability. When a company has an idea, 3D printing can swiftly convert this idea into a tangible prototype. Traditional prototyping often involves lengthy lead times and high costs. Rapid prototyping involves the use of advanced techniques like 3D printing and other forms of digital fabrication to quickly create models or functional parts of telecommunications equipment. This process enables designers and engineers to quickly iterate designs, test functionalities, and refine systems before mass production. With the integration of 3D printing and telecommunications, businesses can easily share digital designs and prototypes with manufacturers, regardless of their geographical location. This enables faster and more cost-effective prototyping. In the fast-paced telecom industry, where technological advancements are constant, rapid prototyping and manufacturing provide a significant competitive advantage. For example, 3D printing enables us to prototype and produce antennas for 5G and 6G networks at a far lower cost. By leveraging 3D printing, engineers are able to design and produce prototypes of complex circuit boards and antennas in-house. Figure 10.5 shows a 3D-printed prototype of a circuit board [2].

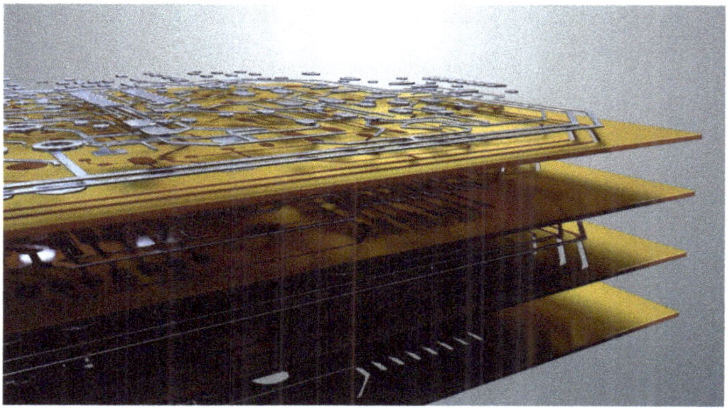

Figure 10.5 A 3D-printed prototype of a circuit board [2].

- *Parts Production:* In addition to prototyping, 3D printing is also used for producing end-use parts in the telecom industry. This is especially true for small parts that are complex to manufacture using traditional manufacturing methods. One example of 3D printing being used for end-use parts in the telecom industry is the production of antenna brackets. Antenna brackets are critical components that hold antennas in place on towers and other structures. Another way in which 3D printing is used in the telecom industry is for creating customized products. With 3D printing, telecom companies can create products that are tailored to the specific needs of individual customers. 3D printing is also being used in the telecom industry for creating tools and fixtures.

- *Remote Production:* Remote production, also known as distributed or decentralized production, refers to the ability to manufacture products in a location separate from where they will be used or consumed. It involves leveraging telecommunications technologies to bridge the gap between designers, manufacturers, and end-users. It has become increasingly relevant in recent times due to advancements in telecommunications and 3D printing technologies. Remote production eliminates the need for traditional supply chains, reducing transportation costs and other associated expenses, and promoting a greener approach to manufacturing. Remote production empowered by telecommunications and 3D printing has the potential to transform industries, making production processes more efficient, cost-effective, and environmentally friendly. The applications of remote production span across various sectors, enabling industries to meet evolving demands, minimize costs, and adapt to a rapidly changing world.

- *Distributed Manufacturing:* This is also known as micro-manufacturing or local fabrication. It is a manufacturing model that utilizes decentralized production facilities to create products. Unlike traditional manufacturing that primarily relies on centralized factories, distributed manufacturing leverages a network of smaller, geographically distributed facilities. By bringing production closer to the end-users, distributed manufacturing enables businesses to meet specific regional demands quickly and efficiently. This approach allows for greater flexibility and customization, as well as reduced lead times and transportation costs. With localized production facilities, businesses can respond rapidly to changing consumer demands and market trends, reducing the time between design and delivery. Eliminating the need for long-distance transportation and inventory storage can significantly reduce costs associated with traditional manufacturing. Distributed manufacturing holds great potential for the future of manufacturing.

- *Satellite Manufacturing:* The integration of 3D printing technology in satellite manufacturing has revolutionized the industry. By leveraging the capabilities of additive manufacturing, satellite engineers can now produce intricate, lightweight components tailored to the specific needs of each mission. This precise and customizable approach not only enhances satellite performance but also opens up possibilities for increased cost efficiency and innovation. As the demand for advanced satellite systems continues to grow, exploring the intersection of 3D printing and satellite manufacturing holds promise for the future of space exploration and communication. The utilization of 3D printing technology in satellite manufacturing presents significant advantages, including the creation of robust and lightweight structures capable of withstanding the harsh conditions of lower earth orbit. 3D printing enables the production of optimized components such as thermal management and radio-frequency parts for satellites. 3D printing in the satellite industry revolutionizes manufacturing processes to meet specific mission requirements efficiently. Figure 10.6 shows satellite production [15].

Figure 10.6 Satellite production [16].

10.6 BENEFITS

3D printing reduces the capital required to achieve economies of scale. It increases flexibility and reduces the capital required to achieve scope. Beyond expanding the universe of designers, software can help expand the universe of AM applications. Hybrid processes combine the benefits of additive and subtractive manufacturing to help improve postprocessing and dimensional and geometric accuracy while reducing production times. Telcos will definitely benefit from lower product cost and faster time to market. Other benefits include the following [13,16]:

- *Customization:* When talking about product customization, 3D printing has a big role in the telecom process. With recent advancements in 3D printing and materials, it is now possible to develop highly customized products with precise shape and accurate dimensions. Patents related to customized industry solutions predominate, targeting diverse industries such as aerospace, automotive, consumer products, medical technology, health care, defense, and industrial products. By leveraging 3D printers, companies can move from mass production to mass customization.

• *Economies of Scope:* Considerations of minimum efficient scale can shape supply chains. AM has the potential to reduce the capital required to reach minimum efficient scale for production, thus lowering the manufacturing barriers to entry for a given location. Economies of scope influence how and what products can be made. Companies take advantage of scale economics offered by AM to help transform supply chains for the products they offer.

• *Flexibility:* Flexibility of AM facilitates an increase in the variety of products a unit of capital can produce, reducing the costs associated with production changeovers and customization and, thus, the overall amount of required capital.

• *Local Production:* With high dependency on specific geographies or central hubs for manufacturing, 3D printing is revitalizing local production by introducing the concept of a decentralized approach. It could potentially enable production close to the customer which would in turn lead to lower distribution cost and allow flexibility to make necessary changes according to domestic attributes. Traditional manufacturing methods often involve transporting goods across countries, leading to high freight costs. However, with 3D printing, localized production becomes a reality, allowing faster delivery to end-users and considerably reducing transport costs.

• *Component Fabrication:* 3D printing offers flexibility in the design, manufacturing, and distribution of complex components as well as helps reduce production wastages. For example, Airbus is exploring 3D printing solution to augment radio frequencies and filter out unwanted signals something that is prone to noise interference. They have adopted metal 3D printing technique to manufacture "waveguides" for telecom satellites. It helps them in terms of mass reduction, accuracy and precision of the components. Whether it is 3D printed wind turbine blades to improve performance, extremely accurate telecom satellites parts, or radio communication on airplanes, 3D printing is making a huge impact.

- *Remote Collaboration:* Telecommunications enables seamless communication across distances, allowing individuals from different locations to collaborate effectively. This is particularly advantageous when combined with 3D printing. Designers, engineers, and experts can work together remotely, sharing 3D models and prototypes in real-time.

- *Distributed Manufacturing:* The combination of 3D printing and telecommunications opens up possibilities for distributed manufacturing. Instead of relying on centralized production facilities, 3D printing allows for the creation of products on-demand, closer to the end-users. This decentralized approach reduces shipping costs and carbon emissions.

- *Cost Savings:* The cost savings associated with 3D printing stem from the ability to optimize material usage and reduce waste, leading to more efficient production processes. The combination of telecom-driven 3D printing and just-in-time manufacturing results in significant cost savings. Manufacturers can minimize wastage by printing only the required parts, reduce transportation costs, and optimize their production processes. For example, the aerospace sector benefits from lowered costs through 3D printing technologies.

10.7 CHALLENGES

Printer providers may consider exploring new fabrication techniques that can address issues such as design complexity, surface finish, unit cost, and speed of operations. They can also consider launching inexpensive yet higher-quality printer models. Other challenges include the following:

- *Standards:* The lack of standardization poses challenges for scaling up the benefits of 3D printing. As with any system that requires interoperability, standards are needed to increase ecosystem interoperability so that designs and data can be more easily shared across different platforms.

- *Confusion:* Many companies remain hesitant about embracing 3D printing, often out of a lack of understanding of how the technology is currently being used in their sector, and uncertainty as to whether it is the right time to invest.

- *Cost:* The cutting edge of additive manufacturing offers a world of possibilities for companies looking to transform their manufacturing processes. But companies that want to tap into that world have traditionally had to invest huge sums of money into the latest 3D printing machines and then figure out how to integrate them into their operations. While the initial investment in 3D printing equipment might be substantial, the long-term benefits and the return on investment (ROI) are profound. Additive manufacturing, the underlying principle of 3D printing, is a game-changer in cost management
- *Material Limitations:* Not all materials are suitable for 3D printing. While a broad range of plastics, metals, and composites are available, some specific materials might not yield the desired results when printed. Additionally, the quality of 3D printing material can significantly affect the outcome.

10.8 CONCLUSION

Although the use of the 3D printing technology is not prevalent in the telecommunications sector, this field is considered as a potential area of growth of the industry, given the increasing use of 3D printed electronics. When combined, 3D printing and telecommunications offer numerous benefits and opportunities across industries. Telecom-driven 3D printing has revolutionized industries worldwide, offering unparalleled opportunities for innovation and optimization. More information about 3D printing can be found in the books in [8,17].

REFERENCE

[1] P. Mota, "An overview of 3D printing in telecom: from ideas to production,"

https://blog.yelco.tech/3d-printing-in-telecom

[2] "How mature is your industry in its adoption of 3D printing? [infographic]," July 2019,

https://amfg.ai/2019/07/09/how-mature-is-your-industry-in-its-adoption-of-3d-printing/

[3] "Communications infrastructure,"

https://www.bv.com/solutions/communications-infrastructure/

[4] M. N. O. Sadiku, P. O. Adebo, and J. O. Sadiku, "3D printing in telecommunication," *International Journal of Trend in Research and Development*, vol. 11, no. 6, November-December 2024, pp. 194-198.

[5] M. N. O. Sadiku, U. C. Chukwu. A. Ajayi-Majebi, and S. M. Musa, "3D printing: An introduction, " *International Journal of Trend in Scientific Research and Development*, vol. 6, no. 7, November-December 2022, pp. 573-577.

[6] F. R. Ishengoma and T. A. B. Mtaho, "3D printing: Developing countries perspectives computer engineering and applications," *International Journal of Computer Applications*, vol. 104, no. 11, October 2014, pp. 30-34.

[7] D. Pitukcharoen, "3D printing booklet for beginners,"

https://www.metmuseum.org/-/media/files/blogs/digital-media/3d printingbookletforbeginners.pdf

[8] S. Torta and J. Torta, *3D Printing: An Introduction. Mercury Learning and Information*, 2019.

[9] T. Hoffman, "3D printing: What you need to know," July 2020,

https://www.pcmag.com/news/3d-printing-what-you-need-to-know

[10] “Beginner’s guide to 3D printing,”

https://www.think3d.in/landing-pages/beginners-guide-to-3d-printing.pdf

[11] “Introduction to 3D printing,”

https://education.gov.mt/en/resources/news/documents/youth%20guarantee/3d%20printing.pdf

[12] “Guide to 3D printing materials: Types, applications, and properties,”

https://formlabs.com/blog/3d-printing-materials/

[13] G. Andersen, “Telecommunications and 3D printing: Enabling distributed manufacturing,” February 2024,

https://moldstud.com/articles/p-telecommunications-and-3d-printing-enabling-distributed-manufacturing#:~:text=Distributed%20Manufacturing%3A%20The%20synergy%20between,closer%20to%20the%20end%2Dusers.

[14] A. Tyrer-Jones, “3D printing ‘world’s largest’ carbon composite rocket on Rocket Lab’s 90-ton 3D printer,” October 2024,

https://3dprintingindustry.com/news/3d-printing-worlds-largest-carbon-composite-rocket-on-rocket-labs-90-ton-3d-printer-233751/

[15] E. Change, “3D printing in satellite manufacturing,” March 2024,

https://telecomworld101.com/3d-printing-satellite-manufacturing/

[16] “Will 3D printing benefit telcos?” April 2020,

https://stl.tech/blog/3d-printing-will-benefit-communication-industry/

[17] M. N. O. Sadiku, *3D Printing and Its Applications*. Moldova,

Europe: Lambert Academic Publishing, 2024.

CHAPTER 11
VIRTUAL AND AUGMENTED REALITY IN TELECOMMUNICATIONS

"Augmented reality will change the world more than a lot of other technologies. Traveling around to meet people will be much less important when you can stand in a room and chat with a virtual representation of a person that's so close to reality – it'll be a whole new level."

—Tim Sweeney

11.1 INTRODUCTION

Augmented reality (AR) and virtual reality (VR) offer innovative ways for marketers to engage and delight customers. Virtual reality is a computer technology that creates a simulated environment, while augmented reality overlays digital information onto real-world objects. VR creates an entirely artificial environment for the user to interact with, while AR is merely adding to the user's real-life experience. Almost any person with a smartphone can get access to augmented reality, making it more efficient than VR as a branding and gaming tool. AR and VR can do amazing things – but only if fueled by the right infrastructure. For AR and VR to work seamlessly, they need a lot of data, and they need it fast [1].

AR uses a real-world setting while VR is completely virtual. AR users can control their presence in the real world; VR users are controlled by the system. VR requires a headset device, but AR can be accessed with a smartphone. AR enhances both the virtual and real world while VR only enhances a fictional reality. Augmented and virtual reality creates new ways of interaction using graphics and gestures that are highly intuitive to human beings. Multiple factors, such as the deployment of a faster network like 5G and the usage of emerging technologies such as AI, have all pushed for greater adoption of AR/VR in telecom and other industries.

Virtual reality (VR) and augmented reality (AR) are immersive technologies that can redefine how we interact with the digital world. They are rapidly transforming various industries, including telecommunications. Their emergence is offering new business opportunities and telecom companies are integrating VR and AR into their services. Telecom companies are leveraging VR devices to create immersive experiences, unlocking opportunities beyond the physical world. VR and AR technologies have the potential to revolutionize the telecommunications industry by enhancing communication, collaboration, and customer experiences while reducing costs and environmental impact [2].

This chapter explores the immersive technologies and their vital roles in telecommunications. It begins with explaining different types of digital realities. It describes virtual reality and augmented reality. It presents the roles of virtual reality and augmented reality in telecommunications. It provides some applications of virtual reality and augmented reality in telecommunications. It highlights the benefits and challenges of virtual reality and augmented reality in telecommunications. The last section concludes with comments.

11.2 TYPES OF DIGITAL REALITIES

The telecommunication industry has become an integral part of today's business world. It is providing uninterrupted services across several businesses and helping them carry out their operations smoothly. It is well known for its complex infrastructure, as typically shown in Figure 11.1 [3].

Figure 11.1 Telecom complex infrastructure [3].

The telecom industry stands out as a primary beneficiary and a crucial driver of VR and AR technologies. For example, VR can provide a simulated environment for remote training and collaboration, while network technicians can use AR to visualize and diagnose problems in complex telecom infrastructure. Thus, integrating VR and AR into telecommunication has the potential to create a more immersive, seamless, and intuitive user experience.

There are four types of digital realities [4,5]:

- *Augmented reality* (AR) — designed to add digital elements over real-world views with limited interaction.
- *Virtual reality* (VR) — immersive experiences helping to isolate users from the real world, usually via a headset device and headphones designed for such activities.
- *Mixed reality* (MR) — combining AR and VR elements so that digital objects can interact with the real world means businesses can design elements anchored within a real environment.
- *Extended reality* (XR) — covering all types of technologies that enhance our senses, including the three types previously mentioned.

These four types are illustrated in Figure 11.2 [6]. AR effectively "augments" or "enhances" our experience of the world around us.

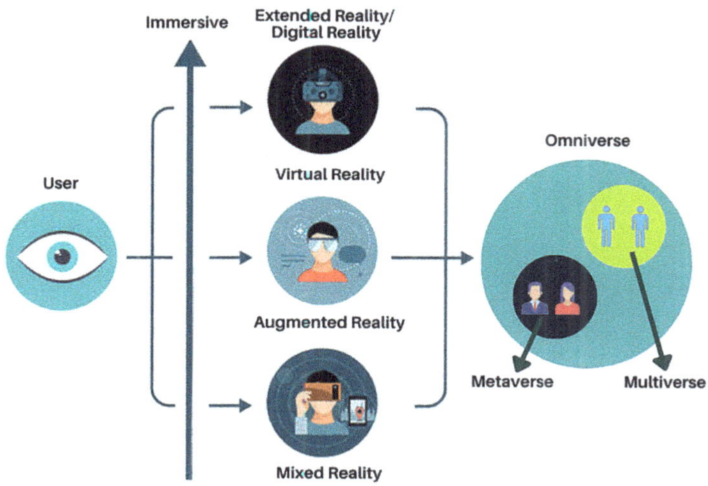

Figure 11.2 Four types of digital realities [6].

11.3 WHAT IS VIRTUAL REALITY?

Human beings are visual creatures. Both hearing and seeing are central to our sense of space. As we have been spending an increasing amount of time in front of computers, the need to explore more of the role of virtual environments also needs to be increased. Virtual reality is the key technology for experiencing sensations of sight, hearing, and touch of the past, present, and future. VR is a fully immersive technology where users wear a head-mounted display and experience a simulated world of imagery and sounds. VR enables active learning [7].

The term "virtual reality" essentially means "near-reality." Virtual reality has been known by different names, such as synthetic environment, cyberspace, artificial reality, virtual environments, and simulator technology. Research into VR dates back to the 1960s when computer scientists developed head-mounted displays for military and medical training purposes. The concept of virtual reality was first introduced by Ivan Sutherland in 1965. His vision was to create an immersive environment that could be used for training, simulation, and entertainment. Since then, VR has come a long way.

The terms, "virtual reality" and "cyberspace" are often used interchangeably. A cyberspace may be regarded as a networked virtual reality. Virtual reality is a simulated experience that can be similar to or different from the real world. It is a computer-generated, 3D environment that completely immerses the senses of sight, sound, and touch. The complete immersion of the senses overwhelms users engrossing them in the action [8]. As shown in Figure 11.3, the mixed reality (MR) continuum captures all possible combinations of the real and virtual worlds [9].

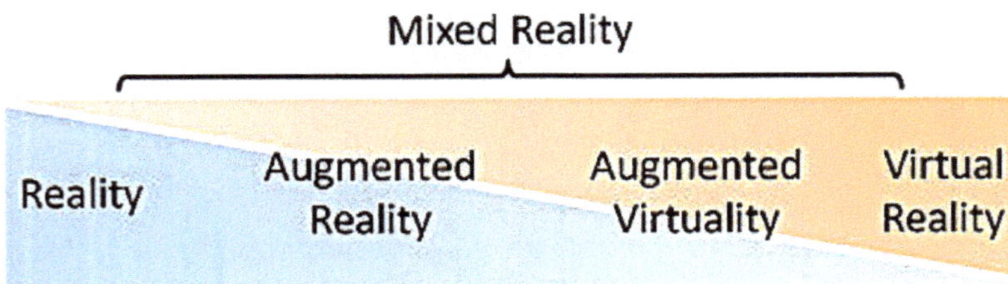

Figure 11.3 The mixed reality continuum captures all possible combinations of the real and virtual worlds [9].

Virtual reality (VR) describes a three-dimensional (3D), computer-generated environment that can be explored and interacted with by a person. Once entered, it becomes a reality to the person. Instead of viewing a screen in front of them as when using traditional interfaces, users are immersed and able to interact with 3D worlds.

Based on data entered by programmers, computers create virtual environments by generating 3D images. Users can view these images by using a head-mounted device, which can be a helmet or goggles. Mobile VR glasses enable users to be present in any environment they want at any time and place. Users may use a joystick or trackball to move through the virtual environment. A person using virtual reality equipment can dive into a fully immersive 3D environment, look around the artificial world, move around in it, and interact with virtual objects. The virtual environment can be viewed using a cell phone screen, monitor, projector, or head-mounted display (HMD). As the person is immersed in the computer-generated environment, the brain is deluded into thinking the virtual world is reality.

VR has three main characteristics [10]: interaction, immersion, and imagination. Interaction refers to the natural interaction between the user and the virtual scene. Immersion means that the user feels that they are part of the virtual world as if they are immersed. Imagination refers to the use of multi-dimensional perception information provided by the VR scenes. A type of VR known as 360VR is most commonly used in education. There are two major ways in which VR is implemented in the classroom: VR headsets and immersive classrooms.

Virtual reality technology includes multiple components divided into two main groups: hardware and software components [11].

- *Hardware Components:* The hardware components include a computer workstation, sensory displays, a tracking system, wearable devices, and input devices. Sensory displays are used to display the simulated virtual worlds to the user. The most common type is the head-mounted displays (HMDs), which is used in combination with tracking systems. Head-mounted displays are shown in Figure 11.4 [12].

Figure 11.4 Head-mounted displays [12].

Users interact with the simulated environment through some wearable devices. VR depends on special responses such as raising hands, turning the head, or swinging the body. A wearable device is important in making these effects realistic. Special input devices are required to interact with the virtual world. These include the 3D mouse, the wired glove, motion controllers, and optical tracking sensors. These devices are used to stimulate our senses together to create the illusion of reality.

- *Software Components:* Besides the hardware, the underlying software plays an important role. It is responsible for the managing of I/O devices and time-critical applications. The software components are 3D modeling software, 2D graphics software, digital sound editing software, and VR simulation software. VR technology has been designed to ensure visual comfort and ergonomic usage.

11.4 WHAT IS AUGMENTED REALITY?

The term "augmented reality" was coined in 1990 by Tom Caudell, a former Boeing researcher. Since then, the technology has vastly improved. The rise in AR use stems from four main developments [13]: (1) the pervasiveness of low-cost visual sensors, such as phone cameras, (2) progress in environmental perception algorithms, such as visual simultaneous localization and mapping, (3) advances in optics, and (4) the maturity of multimedia techniques.

Augmented reality (AR) is a technology that combines real-world environments with computer-generated generated information such as images, text, videos, animations, and sound. It can record and analyze the environment in real-time. The technology is accessible to the ordinary user. It is becoming more attractive as a mainstream technology due to the proliferation of modern mobile computing devices like smartphones and tablet computers with location-based services. For example, AR allows consumers to visualize a product in more detail before they purchase it. This feature enhances consumer interaction and helps them never to repurchase the wrong item.

The key objective of AR is to bring computer-generated objects into the real world and allows the user only to see them. In other words, we use AR to track the position and orientation of the user's head to enhance/augment their perception of the world. Projected images are overlaid on top of a pair of goggles or glasses, which allow the images and interactive virtual objects to lay on top of the user's view of the real world. Thus, augmented/enhanced reality involves extending the real-time environment with a digital overlay, in which computer-generated images and sound enhance real life. The digital overlay scene appears on the actual scene the user is experiencing.

Augmented reality falls into two categories: 2D information overlays and 3D presentations, like those used with games. It combines multiple technologies allowing users to interact with virtual entities in real-time. Various technologies used for augmented reality include a processor, monitors, handheld devices, display systems, sensors, and input devices. Modern mobile computing devices like smartphones and tablet computers contain these elements also, making them suitable AR platforms.

To obtain a sufficiently accurate representation of reality, AR needs the following five components [14]:

- *Sensors:* AR needs suitable sensors in the environment and possibly on a user, including fine-grained geolocation and image recognition. These are activating elements that trigger the display of virtual information.

- *Image Augmentation:* This requires techniques such as image processing and face recognition.

- *Head-mounted Display:* HMDs are used to view the augmented world where the virtual computer-generated information is properly aligned with the real world. Display technologies are of two types: video display and optical see-through display.

- *User Interface:* This includes technologies for input modalities that include gaze tracking, touch, and gesture. AR is a user interface technology in which a camera-recorded view of the real world is augmented with computer-generated content such as graphics, animations, and 2D or 3D models.

- *Information Infrastructure:* AR requires significant computing and communications infrastructure undergirding all these technologies. The infrastructure determines what real-world components to augment, with what, and when.

The AR systems, based on these technologies, should be more accurate, smaller, lighter, faster, simpler, cheaper, and convenient for the users. Google Glass was the first AR platform to get broad public exposure. Figure 11.5 shows the evolution of AR [15].

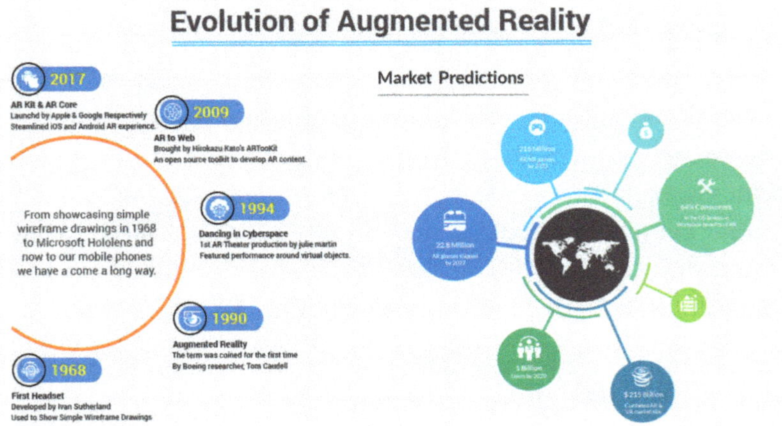

Figure 11.5 The evolution of AR [15].

AR technology has been used in many fields, such as education, healthcare, military, business, engineering, architecture, robotics, manufacturing, entertainment, space industry, maintenance, coding, consumer design, psychology, etc. When incorporating AR in the classroom, it is important to keep the following best practices in mind to ensure a smooth and effective integration [16].

Setting clear goals and objectives for using AR:

- Creating a plan for how AR will be used in the classroom;
- Providing training for educators and students on how to use AR tools;
- Incorporating AR into existing curriculum and lesson plans;
- Assessing and evaluating the effectiveness of AR in achieving educational goals.

11.5 VR AND AR IN TELECOMMUNICATIONS

The world is changing rapidly, and so are the ways in which we communicate. The world of telecommunications is constantly evolving, and AR and VR are at the forefront of this change. Telecommunication companies providing the service for other advanced technological solutions are working hard to keep abreast with the emerging technologies. With the rise of new technologies, we are now seeing a convergence between virtual reality and telecommunication. VR technology is leveraging the capabilities of 5G for improved virtual experiences, ensuring real-time interactions in a virtual environment. VR is paving the way for a new era of AR applications. The advent of these immersive technologies is reshaping telecommunication experiences and playing a vital role in transforming the telecommunication landscape. Telecom companies are utilizing VR to bridge the gap between physical and virtual worlds, creating new opportunities for digital content and business growth [17].

Recent advances in high-speed communication and miniature mobile computing platforms have escalated a strong demand for deeper human-digital interactions beyond traditional flat panel displays. With rapid advances in high-speed communication and computation, augmented reality (AR) and virtual reality (VR) are emerging as next-generation display platforms for deeper human-digital interactions. AR/VR in telecom helps facilitate remote collaboration, support, and training. These technologies will further help telecom companies cut costs, rates, downtimes, and waiting times for services.

11.6 APPLICATIONS OF VR AND AR IN TELECOMMUNICATIONS

VR and AR are gaining significant traction in the telecommunications sector. They are being used to improve everything from training technicians to streamlining customer service in the telecom industry. VR and AR can be integrated into telecommunication in different ways, as typically shown in Figure 11.6 [18]. They are applied in telecommunications in the following ways [17,19]:

Figure 11.6 Ways telcos are leveraging AR/VR [18].

• *Communication:* In the telecommunications industry, soft skills such as customer service, communication, and problem-solving are critical to providing high-quality service and ensuring customer satisfaction. One of the most significant benefits of using VR technology for soft skills training in telecommunications is its ability to simulate real-life scenarios. Virtual reality technology offers an impactful and engaging approach to soft skills training. Traditional forms of communication such as the telephone are being superseded by video conferencing, Skype, and live chat. Sales representatives in the telecommunications industry must have excellent communication and interpersonal skills in order to best represent their company's products and services. VR can facilitate communication between employees and customers. With its ability to provide immersive and interactive experiences, VR is transforming the way companies communicate with their employees, customers, and partners. It can enhance communication, collaboration, and customer experiences while reducing costs and environmental impact. For example, some telecom companies have started using VR to give customers a virtual tour of their facilities. This allows customers to experience the facilities without having to travel, saving time and money.

- *Collaboration:* Collaboration and teamwork are essential skills for employees in the telecommunications industry, as projects often require the involvement of multiple teams and departments. One of the most significant benefits of VR in telecommunications is its ability to improve collaboration between geographically dispersed teams. Telecom companies are exploring the use of VR for remote collaboration. With VR, team members can collaborate in a virtual environment, regardless of their physical location. This allows teams to work together in real-time, access shared files, and collaborate seamlessly, allowing them to complete tasks more efficiently. This can help to improve productivity and reduce travel costs.

- *Healthcare:* In recent years, VR telecommunication has revolutionized healthcare by enabling remote surgeries and medical training. It enhances patient therapy and rehabilitation and enables medical consultations through virtual meetings. Telecom companies use VR for remote collaboration in healthcare research, improving customer experience. The integration of VR devices in the healthcare sector has created a virtual environment that bridges the gap between the real world and the physical world, resulting in real-time connectivity for medical professionals. Telecommunications can be used to help virtual reality systems such as surgery simulation or telemedicine. An example of this is remote surgery in which images from that surgery can be transmitted to various locations around the world. It also enables surgery to be performed in remote locations using robotic technology and virtual reality.

- *Interactive Customer Service:* The most visible impact of VR and AR in telecom is on the customer experience. Telecom businesses are using these technologies to offer immersive product demonstrations and interactive customer service experiences. Customers might face issues with their telecom services or equipment and might not be able to describe them clearly over the phone or may not be able to follow the instructions/steps given over calls for troubleshooting the issues. This hampers customer experience and creates misunderstanding. AR can revolutionize customer service in the telecom industry. Customers through their AR-enabled smart devices can share live videos of the issue with the technicians or the customer service and support representatives. With AR, customer service reps can see exactly what you see through your smartphone camera. This allows them to visually identify a problem and guide you through the troubleshooting process in real time. This eliminates the need for lengthy explanations and frustrating guesswork, resulting in a faster and more satisfying customer experience. Customer service representatives can practice handling different customer interactions, such as complaints, inquiries, and technical support.

- *Business Intelligence:* The traditional methods of data representation and analysis may not provide a comprehensive view or understanding. This could be a challenge while gaining data insights and business intelligence for more informed decisions. Using VR headsets, complex business information/data can be presented in a simple manner for ease of understanding. Using business analytics, companies could understand the current market dynamics as well as predict future trends and make smart decisions. VR in telecom helps businesses to view data analytics in an easier, more comprehensive manner and create strategies that hit the bull's eye.

- *Augmented Maintenance:* VR and AR are also revolutionizing network management and maintenance. Keeping telecom networks up and running requires a team of skilled maintenance technicians. VR in telecom is capable of creating simulated environments. Thus, it can simulate the installation and maintenance of network equipment as well. AR can empower these technicians by giving them a real-time view of the equipment they are working on. This allows them to identify problems more quickly and efficiently and make repairs more accurately. AR can also be used to remotely guide technicians through complex repair procedures, reducing the need for on-site supervision and improving overall maintenance efficiency. The AR headsets will display the equipment manual, specifications, as well as maintenance instructions, the last inspection date, the history of repairs, etc. This helps the technicians understand the equipment better and schedule and perform their routine maintenance tasks.

- *Virtual Conferences:* The pandemic has made virtual conferences and events common globally. The costs for physical arrangements are reduced in comparison to traditional conferences. In addition, participants can attend these events from their homes or on the go, eliminating commutation costs. Participants could interact with each other in a simulated environment; creating excitement and engagement. They save travel costs, especially for conferences or events at international locations, where a participant may face visa issues, budget constraints, etc.

- *Field Service Management:* Field technicians face a plethora of issues daily since they must oversee complex networks and equipment. They further have to offer solutions for issues that vary in complexity and repair faults on the field. This is when field technicians take the help of AR technology. Through AR in telecom, field technicians could identify issues or defects and ascertain the time required for repairs. Using this technology field technicians could also identify different components of a machine.

- *Audits and Record-keeping:* In the telecom industry, conducting site audits the traditional way can be quite time-consuming and resource intensive. By using AR headsets, telecom companies can conduct regular audits and ensure operational efficiency and security. AR technology allows for performance analysis of the equipment. Field technicians using AR devices like smart glasses or smartphones too can access the data, documentation, as well as checklists in real-time while conducting on-site audits.

11.7 BENEFITS

With advancements in technology, VR has become more affordable and accessible, making it possible for anyone to experience VR. Virtual reality has enormous potential in enhancing communication, collaboration, and customer experiences in telecommunications. Augmented reality in telecom allows companies to offer more interactive and effective training sessions and ensure proper maintenance of the equipment to avoid any potential accidents or failures. Other benefits include [18,20,21]:

- *Enhanced Collaboration:* AR/VR in telecom facilitates better remote communication and collaboration. Two-way video and audio sessions powered by AR bring field technicians or professionals and the equipment experts closer; allowing the latter to view and monitor the equipment in real-time. This is especially helpful in the case of any issues or faults in the equipment. With the help of a virtual, or simulated environment, employees working from different parts of the city, country, or world could collaborate seamlessly.

- *Remote Work:* VR technology can create a virtual workspace, allowing employees to attend meetings, conferences, and training sessions from the comfort of their home. This can be particularly useful for companies with employees in different time zones or locations. With VR, employees can interact with each other and their environment in real-time, making remote work more engaging and productive. For example, with AR, technicians can receive real-time guidance from experts located anywhere in the world.

- *Training:* VR can provide training opportunities for employees. For example, telecom companies can use VR to simulate real-world scenarios, allowing employees to practice and improve their skills in a safe and controlled environment. This can be particularly useful for employees who work in hazardous or complex environments. Leadership skills are essential for executives, managers, and supervisors in the telecommunications industry. VR technology can be used to simulate a variety of leadership training scenarios. VR simulations can train field engineers to safely climb and work on cellular towers, requiring extensive safety knowledge and experience. VR training reduces equipment downtime, as trainees do not require live equipment for practice, making it a cost-effective and efficient solution for telecom training programs.

- *Enhanced Customer Experiences:* Telecom companies can provide their customers with equipment manuals that are immersive and engaging as well as informative. These instruction manuals make it easier for the customers to learn and maintain the equipment. With VR and AR, customers could be joined by a professional or an expert whenever required. They could walk the customers through the process of fixing common issues or assembling the equipment and save the time and costs spent on traveling to physical locations for support.

- *Diversity Training:* Diversity and inclusion are critical issues in the telecommunications industry, as the workforce and customer base become more diverse. VR technology can be used to provide employees with a safe and immersive environment to explore issues related to diversity and inclusion. For example, VR simulations can be used to explore scenarios related to unconscious bias, cultural differences, and workplace harassment.

11.8 CHALLENGES

While AR/VR technology offers exciting possibilities for the telecom industry, some challenges still need to be addressed. The biggest challenge facing VR is its technical constraints and hardware limitations. The challenges faced in integrating VR and AR into telecommunication include technical hurdles, user resistance to these new technologies, remote connectivity, and accessibility. A main challenge is that of dealing with a medium which mainly relies upon tone of voice, intonation, gesture and body language as compared to spoken words. Another significant technical hurdle is bandwidth. Other challenges include the following [16]:

- *Privacy and Security Concerns:* A significant challenge facing VR is privacy and security concerns. With VR, users are creating a vast amount of personal data, including body movements, facial expressions, and voice commands. This data can be vulnerable to security breaches, and users should be aware of these risks.

• *Accessibility:* Another limitation of VR is accessibility and inclusivity issues. VR devices can be expensive and require additional hardware, such as powerful graphics cards, which can exclude individuals with disabilities or those with limited financial resources.

• *Standards and Regulations:* With the rapid advancement of VR, there is a need for industry standards and regulations to ensure that VR devices and content are safe and secure for users. Standards and regulations can help create a more stable and trustworthy environment for VR, encouraging further adoption. Standardization is still a work in progress, with various stakeholders involved.

• *Interoperability:* There are also interoperability challenges. For VR and AR to reach their full potential, there needs to be a standardized platform where different devices and applications can interact seamlessly.

• *Latency:* Virtual reality technology encounters latency and bandwidth issues in telecommunication applications.

• *User Resistance:* To address user concerns, clear and concise instructions for using VR telecommunications tools are vital. Providing in-depth training, incorporating user feedback, and emphasizing benefits compared to traditional methods can help overcome resistance.

• *Cost:* One challenge is the cost of AR/VR headsets, which can be a barrier to adoption for some companies. However, as the technology matures, the cost is expected to decrease.

11.9 CONCLUSION

The telecommunications industry has seen a gradual adoption of VR, with companies exploring its potential for enhancing communication and collaboration among teams. VR telecommunication is revolutionizing the way we communicate and interact. The implementation of AR/VR in telecom has created a more productive, and efficient working environment for the workers and enhanced experience for the customers.

While a growing number of telecom companies have started exploring AR/VR technology for multiple use cases, the ones reluctant to explore and embrace it might be left behind in the marketplace. As virtual and augmented reality become more entwined in how we work, play, and learn, the industry will only continue to grow. As the technologies become mature, we will see greater adoption of AR/VR in telecom. The future of VR/AR telecommunication looks promising, with advancements making it more immersive and interactive.

REFERENCE

[1] "What's the difference between AR and VR?"

https://sopa.tulane.edu/blog/whats-difference-between-ar-and-vr#:~:text=AR%20uses%20a%20real%2Dworld,only%20enhances%20a%20fictional%20reality

[2] M. N. O. Sadiku, P. O. Adebo, and J. O. Sadiku, "Virtual and augmented reality in telecommunications," *International Journal of Trend in Research and Development*, vol. 11, no. 6, November-December 2024, pp. 187-193.

[3] "The role of telecommunications in modern business success,"

https://bizinp.com/article/telecommunications-business-success

[4] M. N. O. Sadiku, C. M. M. Kotteti, and S. M. Musa, "Augmented reality: A primer," *International Journal of Trend in Research and Development*, vol. 7, no. 3, 2020.

[5] "What is augmented reality or AR?"

https://dynamics.microsoft.com/en-us/mixed-reality/guides/what-is-augmented-reality-ar/

[6] "Metaverse and multiverse: The real sense of AR, VR, MR, XR and IR,"

https://yp.ieee.org/metaverse-and-multiverse-the-real-sense-of-ar-vr-mr-xr-and-ir/

[7] M. N. O. Sadiku, J. O. Sadiku, and U. C. Chukwu, "Virtual reality in education," *International Journal on Integrated Education*, vol. 27, vol. 6, no. 3, March 2023.

[8] M. N. O. Sadiku, K. G. Eze, and S. M. Musa, "Virtual reality: A primer," *International Journal of Trend in Research and Development*, vol. 7, no. 2, March-April 2020, pp. 160-162.

[9] "Introduction to augmented reality," June 2016,

https://www.informit.com/articles/article.

aspx?p=2516729&seqNum=4

[10] Y. Li et al., "Gesture interaction in virtual reality," *Virtual Reality & Intelligent Hardware*, vol.1, no.1, February 2019, pp. 84-112.

[11] M. O. Onyesolu and F. U. Eze, "Understanding virtual reality technology: Advances and applications," *Advances in Computer Science and Engineering*, March 2011, pp. 53-70.

[12] "VR rundown: What you need to know before buying a VR System for your school,"

https://vreddo.com.au/vr-rundown-what-you-need-to-know-before-buying-a-vr-system-for-your-school/

[13] H. Ling, "Augmented reality in reality," *IEEE MultiMedia*, July–September 2017, pp. 10-15.

[14] M. Singh and M. P. Singh, "Augmented reality interfaces," *IEEE Internet Computing*, November/December 2013, pp. 66-70.

[15] D. Miller, "The potential of augmented reality in business," December 2018,

https://www.epikso.com/blog/the-potential-of-augmented-reality-in-business/

[16] N. Sahota, "Augmented reality in education: The future of learning is here," February 2023,

https://www.neilsahota.com/augmented-reality-in-education-the-future-of-learning-is-here/

[17] D. Kliuch, "VR in telecommunication: Exploring futuristic trends," May 2024,

https://whimsygames.co/blog/vr-in-telecommunication-exploring-futuristic-trends/#:~:text=Telecom%20companies%20utilize%20VR%20simulations,industries%20with%20its%20innovative%20applications.

[18] "Telcos get techy: How AR/VR is revolutionizing telecom,"

https://www.linkedin.com/pulse/telcos-get-techy-how-arvr-revolutionizing-telecom-maplworld-sbylc/

[19] H. Arora, "AR/VR in Telecom – Use Cases & Benefits,"

https://bigohtech.com/ar-vr-in-telecom/

[20] "Embracing the future: Unveiling the potential of virtual reality in telecommunications," June 2023,

https://www.talespin.com/reading/embracing-the-future-unveiling-the-potential-of-virtual-reality-in-telecommunications

[21] "Virtual reality telecommunications training use cases," March 2023,

https://www.talespin.com/blog-post/virtual-reality-telecommunications-training-use-cases

CHAPTER 12
CYBERSECURITY IN TELECOMMUNICATIONS

"The goal of cybersecurity is not to eliminate all risks, but to intelligently manage risks to an acceptable level."

—Michael Coates

12.1 INTRODUCTION

Telecommunication forms the backbone of nearly every other critical infrastructure, supporting sectors like healthcare, finance, transportation, and government operations. The infrastructure that underpins modern telecommunications companies is typically large and complex. From satellite companies, Internet providers, telephone corporations, the infrastructure behind these organizations makes it feasible for all our videos, audio, and text to be sent around the globe. The infrastructural complexity is typically shown in Figure 12.1 [1]. The expansive infrastructure sees constant change as technology progresses.

Figure 12.1 Telecommunication infrastructural complexity [1].

The telecommunications sector keeps the world connected, as shown in Figure 12.2 [2]. From private communications to business interactions, it is an intrinsic part of our daily lives.

Figure 12.2 Telecommunications sector keeps the world connected [2].

Telecom networks form the backbone of global communication systems, making them lucrative targets for cyber attacks. The telecommunications sector faces a variety of cyber threats that can compromise individual privacy, corporate security, and national safety. Telecom operators store various data, from social security numbers to credit card details. This wealth of sensitive data makes large-scale telecom firms appealing targets for bad actors. These risks underscore the growing need for reinforcing robust cybersecurity measures within the telecom sector.

Industries are facing an exponential increase in cyber threats, which are becoming more complex and sophisticated as the population and organizations become increasingly more digital. Cybersecurity refers to the protection of a computer system and/or network from attack as well as the risk of losing data or information, unauthorized access, and fraud. It entails the adequate use of tools, measures, and strategies to prevent or lessen the impact of cyberattacks. Cybersecurity is becoming increasingly significant due to the increased reliance on computer systems, the Internet, and wireless network standards. A lack of cybersecurity can result in the unavailability of essential industries, such as health and emergency services, which rely on telecommunications to transmit information and provide support to the public. Any compromise in telecom security can hinder urgent communication and coordination efforts during critical times [3].

In this chapter, we explore how cybersecurity measures are applied in the telecommunications industry. It begins with explaining what cybersecurity is all about. It discusses telecom cybersecurity and provides some of its applications. It highlights the benefits and challenges of telecom cybersecurity. The last section concludes with some comments.

12.2 OVERVIEW ON CYBERSECURITY

Cybersecurity refers to a set of technologies and practices designed to protect networks and information from damage or unauthorized access. It is vital because governments, companies, and military organizations collect, process, and store a lot of data. As shown in Figure 12.3, cybersecurity involves multiple issues related to people, process, and technology [4]. Figure 12.4 shows different components of cybersecurity [5].

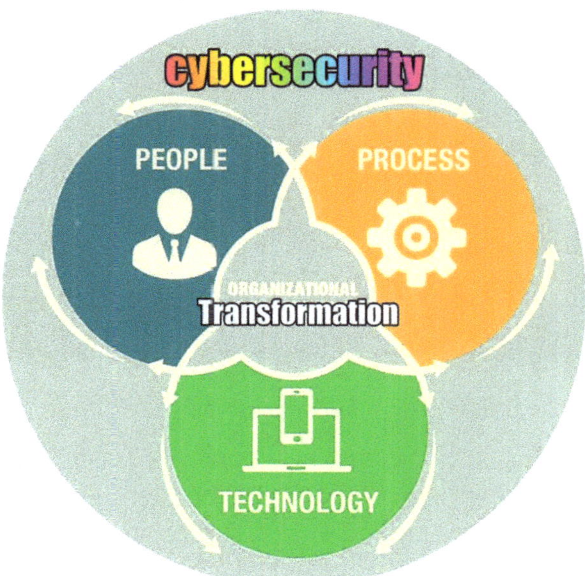

Figure 12.3 Cybersecurity involves multiple issues related to people, process, and technology [4].

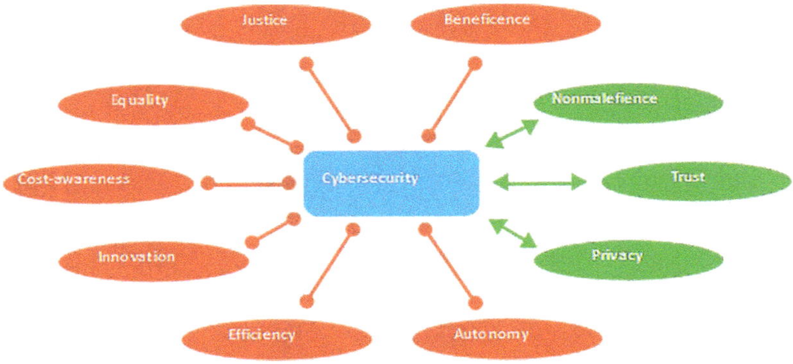

Figure 12.4 Different components of cybersecurity [5] (Green: supportive; red: in tension).

A typical cyber attack is an attempt by adversaries or cybercriminals to gain access to and modify their target's computer system or network. Cybercriminals or ethical hackers are modern-day digital warriors, possessing extraordinary skills and knowledge to breach even the most impregnable systems. A typical cybercriminal is shown on Figure 12.5 [6].

Figure 12.5 A typical cybercriminal [6].

Cyber attacks are becoming more frequent, sophisticated, dangerous, and destructive. They are threatening the operation of businesses, banks, companies, and government networks. They vary from illegal crime of individual citizen (hacking) to actions of groups (terrorists) [7].

The cybersecurity is a dynamic, interdisciplinary field involving information systems, computer science, and criminology. The security objectives have been availability, authentication, confidentiality, nonrepudiation, and integrity. A security incident is an act that threatens the confidentiality, integrity, or availability of information assets and systems [7].

- *Availability:* This refers to availability of information and ensuring that authorized parties can access the information when needed. Attacks targeting availability of service generally leads to denial of service.
- *Authenticity:* This ensures that the identity of an individual user or system is the identity claimed. This usually involves using username and password to validate the identity of the user. It may also take the form of what you have such as a driver's license, an RSA token, or a smart card.
- *Integrity:* Data integrity means information is authentic and complete. This assures that data, devices, and processes are free from tampering. Data should be free from injection, deletion, or corruption. When integrity is targeted, nonrepudiation is also affected.
- *Confidentiality:* Confidentiality ensures that measures are taken to prevent sensitive information from reaching the wrong persons. Data secrecy is important especially for privacy-sensitive data such as user personal information and meter readings.
- *Nonrepudiation:* This is an assurance of the responsibility to an action. The source should not be able to deny having sent a message, while the destination should not deny having received it. This security objective is essential for accountability and liability.

Everybody is at risk for a cyber attack. Cyber attacks vary from illegal crime of individual citizen (hacking) to actions of groups (terrorists). The following are typical examples of cyber attacks or threats [9]:

- *Malware:* This is a malicious software or code that includes traditional computer viruses, computer worms, and Trojan horse programs. Malware can infiltrate your network through the Internet, downloads, attachments, email, social media, and other platforms. Spyware is a type of malware that collects information without the victim's knowledge.

- *Phishing:* Criminals trick victims into handing over their personal information such as online passwords, social security number, and credit card numbers.
- *Denial-of-Service Attacks:* These are designed to make a network resource unavailable to its intended users. These can prevent the user from accessing email, websites, online accounts or other services.
- *Social Engineering Attacks:* A cyber criminal attempts to trick users to disclose sensitive information. A social engineer aims to convince a user through impersonation to disclose secrets such as passwords, card numbers, or social security number.
- *Man-In-the-Middle Attack:* This is a cyber attack where a malicious attacker secretly inserts him/herself into a conversation between two parties who believe they are directly communicating with each other. A common example of man-in-the-middle attacks is eavesdropping. The goal of such an attack is to steal personal information.

These and other cyber attacks or threats are shown in Figure 12.6 [10]. Sources of cyber threats are displayed in Figure 12.7 [11].

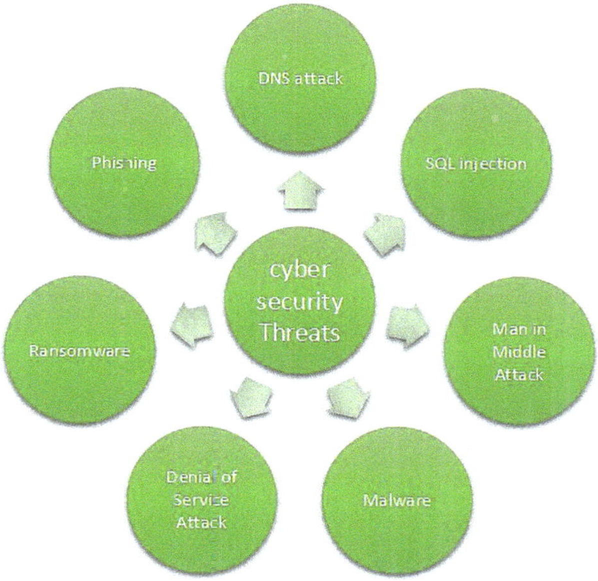

Figure 12.6 Common types of cybersecurity threats [10].

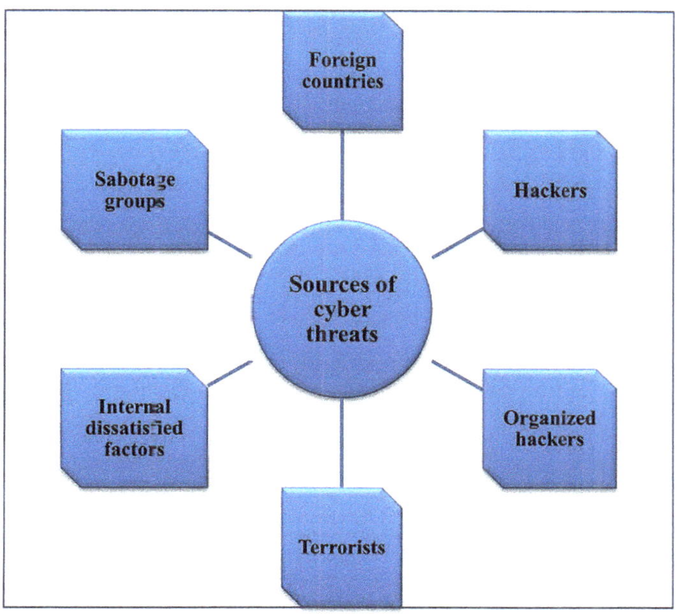

Figure 12.7 Sources of cyber threats [11].

The social and financial importance of cybersecurity is

increasingly being recognized by businesses, organizations, and governments. Cybersecurity involves reducing the risk of cyber attacks. Cyber risks should be managed proactively by the management. Cybersecurity technologies such as firewalls are widely available [12]. Cybersecurity is the joint responsibility of all relevant stakeholders including government, business, infrastructure owners, and users. Cybersecurity experts have shown that passwords are highly vulnerable to cyber threats, compromising personal data, credit card records, and even social security numbers. Governments and international organizations play a key role in cybersecurity issues. Securing the cyberspace is of high priority to the US Department of Homeland Security (DHS). Vendors that offer mobile security solutions include Zimperium, MobileIron Skycure, Lookout, and Wandera.

12.3 TELECOM CYBERSECURITY

Telecom infrastructure faces both physical and cyber vulnerabilities. Physical vulnerabilities include infrastructure location and exposure, making assets susceptible to natural disasters or physical attacks. Cyber vulnerabilities pose significant risks to data integrity and confidentiality, including threats to sensitive information and potential unauthorized access points. The key objective of any cybersecurity program is ensuring the confidentiality of sensitive information, upholding data integrity, and guaranteeing the availability of critical resources.

Telecom cybersecurity refers to the measures and technologies employed to protect telecommunications systems from unauthorized access, attack, or damage. It encompasses the protection of data flowing through mobile devices, telecommunication networks, and other communication infrastructures. Here are some of the common cyber threats in telecom cybersecurity [13]:

- Distributed denial of service (DDoS) attacks
- Data breaches
- Man-in-the-middle attacks
- Ransomware attacks

- 5G infrastructure exploits
- Internet of things (IoT) security threats
- Social engineering attacks
- Advanced persistent threats
- Terrorism and nation-state actors

Some of these cybersecurity threats are depicted in Figure 12.8 [2].

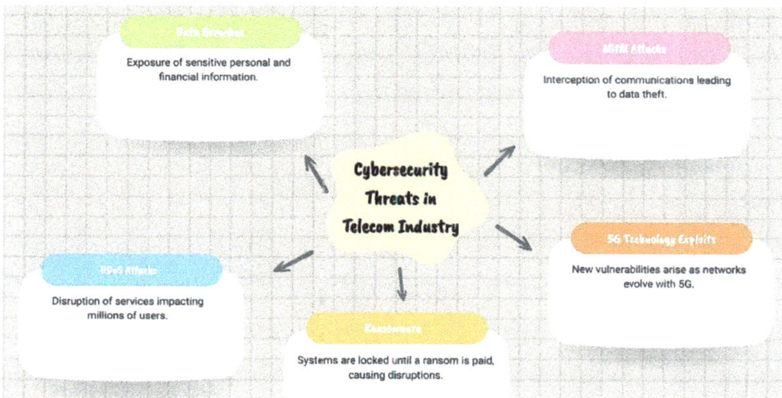

Figure 12.8 Cybersecurity threats in telecom industry [2].

These vulnerabilities make the telecom sector an attractive target for cybercriminals. As a result, robust protection against cyber breaches and preventative cybersecurity solutions are fundamental to the integrity and continuity of operations in the telecom industry. The responsibility to fend off these attacks lies equally on individuals, businesses, and, most significantly, the experts in the field of cybersecurity.

Telecom cybersecurity focuses on safeguarding of telecom infrastructure such as servers, data centers, and network equipment, as well as the software used for managing and transmitting communication data. It also focuses on protecting the vast amounts of sensitive data transmitted across these networks, including personal user data, business communication, and government information. Protecting the sensitive data of large customer bases, which includes personal and financial information, is of legal concern and imperative.

12.4 APPLICATION OF TELECOM CYBERSECURITY

The telecommunication sector is constantly expanding and innovating, making the implementation of continuous monitoring systems essential to meet cyber security challenges. It handles essential infrastructure, therefore a cyber attack can have a huge and far-reaching impact. For many years telecommunication companies have been a prime target for cybercriminals and nation-state actors, due to their role in managing crucial communication networks that handle vast quantities of private and confidential information. The telecommunications industry is facing unprecedented security challenges due to the emergence of new technologies and services. It implements cybersecurity solutions into its operations in the following ways [14]:

- *Communication Networks:* The telecom industry serves as the primary conduit through which information flows across the globe. Every time you make a phone call, send a text, browse the Internet or even use a smart device, you are utilizing the communication networks and infrastructure services provided by the telecommunications industry. Telecommunication infrastructure connects villages, cities, nations, and continents through communication networks, which can be hardwired or wireless and they depend on rural areas, urban areas, or remote areas. Telecommunications infrastructure assets include mobile technologies such as telephone wires, satellites, cables, microwaves, fifth-generation (5G) mobile networks, etc. These telecoms networks are also used to build, control, and operate other critical infrastructure sectors, including energy, information technology, and transportation systems. Amidst this expansive network of communication, the paramount concern is ensuring the security and integrity of these connections. The increasing complexity and interconnectivity of communication networks have led to vulnerability in communication networks. Communication networks need to be resilient. In essence, telecom security aims to ensure the confidentiality, integrity, and availability of communication networks. As technology evolves, so do the challenges and opportunities in securing communication networks.

- *Internet of Things:* IoT adoption has been on the rise over recent years. With an increasing number of devices connected to the network, the threat surface is increasing. In 2021, Gartner estimates that some 25 billion IoT devices will be connected to telecom networks. Accommodating such an increased volume of data is just one part of the challenge for telecom security though. Some of the major risks associated with IoT include system vulnerabilities and weak passwords.

- *5G Network:* The telecom sector has transformed significantly and with the 5th Generation (5G), it is seeing a rapid shift in the services. Cybersecurity concerns surrounding 5G are multifaceted. Some are architectural, pertaining to the novel ways networks are constructed, while others are about the volume and speed of data being transferred. The diverse range of devices poised to harness 5G is staggering. Every connected device is a potential breach point, from smartphones to automated vehicles. The threats are not limited to personal data theft; they range from disrupting vital services to coordinated assaults on urban infrastructure. The interconnectedness that 5G promotes is its strength and its potential weakness. As 5G seeks to unify previously disparate networks, a single vulnerability could snowball into systemic failures. Addressing these vulnerabilities requires a holistic strategy.

- Customer Data: This is another classic high-impact target. Telecom firms commonly keep personal data about all of their customers, such as names, addresses, and even financial information. From billing information to call logs, telecom companies have access to a wealth of sensitive customer data. Due to the vast amounts of personal information telcos gather and keep on their clients, organized cybercriminal gangs also view telecom companies as high-value targets. Cybercriminals or insiders looking to extort clients and steal money will find this confidential information to be a seductive target. Cyberattacks, unauthorized access, and data breaches can result in severe consequences, including financial losses, reputational damage, and national security risks.

12.5 BENEFITS

The importance of cybersecurity in telecommunications cannot be overstated. Given the sensitive nature of the telecom industry, it is critical for telecom companies to prioritize cybersecurity. Prioritizing cybersecurity contributes to stable business operations and opens new opportunities while keeping users safe and protected. Cybersecurity is necessary because security measures protect all forms of data from loss, cyber risk, and identity theft. It is critical because it safeguards all types of data against theft and loss. Other benefits include the following [13]:

- *Low Latency:* The reduced latency in 5G improves the effectiveness of security measures by enabling real-time security monitoring and quicker responses to potential threats, enhancing overall 5G cybersecurity.

- *Protecting Customer Data:* Telecom companies handle vast amounts of sensitive customer data, including personal information, financial data, and communication records. Cyber attacks on telecom companies can result in the theft or compromise of this data, which can be a significant breach of privacy and security.

- *Ensuring Network Security:* Telecom companies provide critical network infrastructure, including voice and data transmission, Internet connectivity, and wireless services. These networks are vulnerable to cyber attacks, which can disrupt services, compromise data, and affect business operations. Cybersecurity training can help telecom companies build and maintain secure networks that are resilient to cyber threats.

- *Compliance with Regulations:* Telecom companies are subject to a range of regulations and standards related to data privacy and security. Compliance with these regulations requires a robust cybersecurity program, which includes regular training for employees.

- *Reputational Risk:* When unauthorized users gain access to telecommunications infrastructure, they can disrupt service and steal personal information. Cyber attacks can have significant financial and reputational consequences for telecom companies. The cost of a cyber attack can include lost revenue, legal fees, and damage to the company's reputation. Telecommunications operators, third-party providers, and subscribers of telecommunications services are at risk. Cybersecurity training can help mitigate these risks by ensuring that employees are equipped with the knowledge and skills necessary to identify and respond to cyber threats.

12.6 CHALLENGES

Protecting telecom infrastructure is far from easy. The telecom industry is facing an uphill battle when it comes to cybersecurity. Attacks are becoming more sophisticated and skilled individuals are able to infiltrate telecom providers. The industry understands that no threat can be tackled in isolation, and that threat actors will continue to exploit vulnerabilities in adopted technologies to achieve their goals. Among the challenges faced by the telecommunication industry, digital security failures are considered risk factors for the operation of services. Preventing unauthorized access, securing data transmissions and ensuring smooth monitoring of a much larger attack surface are the key security challenges for telcos. These failures expose sensitive data and threaten business continuity. Other challenges include [15,16]:

- *Data Security:* Confidential information is always at stake in the growing phase of hackers and malware. Data security offers a shield to the information, protecting it from possible threats. Cybersecurity helps protect customer data, intellectual property, and financial records. It also reduces the possibility of data breaches.

- *Supply Chain Vulnerabilities:* The telecom sector deals with multiple third-party entities such as vendors, web hosting services, data management services, managed service providers, partners, etc. Telecom companies are often the target of "supply chain attacks." Telecom operators often turn to third-party vendors for the infrastructure to support their core services. However, any vulnerabilities within these external entities can have a cascading effect on the entire supply chain.

- *DDoS Attacks:* Distributed denial of service (DDoS) attacks are prevalent in the telecommunications sector. These attacks overwhelm networks with a flood of Internet traffic, rendering them inoperable and denying service to legitimate users. A DDoS attack is instigated by a multitude of malware-infected host machines controlled by an attacker. In telecommunications, where operational continuity is critical, DDoS attacks can lead to a standstill in services, impacting millions of users who rely on uninterrupted connectivity.

- *Phishing:* Phishing emails are gateways to ransomware attacks. Unsuspecting personnel in telecom companies become the focal point of phishing emails designed to lure victims into clicking links or downloading attachments laced with malicious software.

- *Insider Threats:* These are one of the major risks for the telecom industry. When authorized users such as employees, business stakeholders, or independent contractors abuse their legitimate access deliberately or unintentionally, it gives rise to an insider threat. Insiders in the telecom sector can potentially compromise subscriber data, duplicate SIM cards, and more. An issue within telecom is that many employees/insiders are completely unaware that they are a threat in the first place. A lack of understanding or a single misstep can expose gaping holes ready for exploitation.

- *Cost:* This is also a contributing factor, as many organizations have limited resources, and are unable to secure their devices, systems, people, and processes internally.

- *Awareness:* One of the most important things telecom companies can do when it comes to cybersecurity is to educate their employees about best practices and common threats. Many attacks are successful simply because employees are unaware of how to identify or avoid them. Regular training on phishing scams and website security can empower employees to be part of the organization's defense against cyberattacks.

- *Collaboration:* This is vital in the ever-evolving landscape of cybersecurity. Cybersecurity is not a standalone field. By leveraging collective knowledge and experience, we can build a resilient ecosystem that protects critical infrastructure. Engaging with knowledgeable professionals provides valuable insights and best practices for vulnerability assessment. We can also safeguard sensitive data, ensuring seamless communication flow.

- *Regulatory Compliance:* Compliance with regulatory requirements is crucial for minimizing vulnerabilities within the telecom sector. Adhering to regulations and best practices helps protect networks from cyberattacks and unauthorized access, instilling consumer trust in the industry.

- *Standards:* Compliance requirements within the telecom industry revolve around adhering to specific standards and regulations to ensure the security and reliability of communication networks. The National Institute of Standards and Technology [NIST] framework offers a comprehensive set of guidelines, standards, and best practices for improving critical infrastructure cybersecurity. Telecom companies leverage NIST's framework to enhance their cybersecurity posture. The Telecommunications Security Act [TSA] sets out regulations and standards to ensure the security and integrity of telecommunication networks. Compared to other frameworks, TSA offers sector-specific guidelines tailored to the telecom industry's unique challenges. In the United States, the FCC plays a crucial role in regulating communication services and ensuring compliance with security standards. The International Telecommunication Union (ITU) develops global standards for telecommunications and assists countries in implementing secure communication networks.

- *Legacy Technology:* The telecommunications sector still uses legacy technology which makes it vulnerable to IP-based threats. The adoption and transition from legacy systems has been slow.

- *Threat Intelligence:* In the face of sophisticated cyber threats, telecom threat intelligence (TI) is becoming increasingly important. TI in the telecom sector includes information about potential threats, methods used by cyber attackers, and predictions about future cyber threats. This specialized intelligence is crucial for understanding and combating the unique threats faced by the telecom sector, which plays a critical role in global communication.

- *Human Intelligence:* In the complex landscape of telecom cybersecurity, while technology is a critical tool, the real strength lies in human intelligence. The effectiveness of cybersecurity strategies largely depends on the skill, insight, and adaptability of the security teams, along with the broader organizational culture of security. The combination of human intelligence with technological tools forms the backbone of effective cybersecurity in telecom. The true fortification of cybersecurity in the telecom sector lies in a harmonious blend of human intelligence and technology.

12.7 CONCLUSION

The modern telecommunications ecosystem is a complex beast. Its tentacles reach far and wide, affecting everything from mundane daily tasks to global business operations. Our economies and entire business infrastructures are built on modern telecoms. It is a high-value target for cybercriminals due to its extensive use of critical infrastructure and the large amounts of sensitive information it handles. This era demands a dynamic and adaptable approach to cybersecurity, one that evolves in tandem with the rapidly changing technology landscape. Due to the fact that telecommunications companies manage critical infrastructure, a cyberattack might have a significant and wide-ranging effect.

The future of telecom cybersecurity is intertwined with innovation and collaboration. It lies in leveraging advancements in AI-driven threat detection, implementing robust encryption standards and embracing a holistic approach that integrates security into every aspect of network architecture. By prioritizing cybersecurity, telecom companies will not only safeguard their networks but also reinforce trust, enabling a future where seamless, secure communication remains a cornerstone of our interconnected world. More information on the implementation of 5G networks in the telecommunications industry is available from the books in [17-21].

REFERENCE

[1] "Security challenges for the telecom industry 2022,"

https://www.lockmanage.com/security-challenges-for-the-telecom-industry-2022/

[2] "Telecom under siege: A list of cybersecurity threats," Unknown source.

[3] M. N. O. Sadiku, P. A. Adekunte, and J. O. Sadiku, "Cybersecurity in telecommunications," *International Journal of Trend in Scientific Research and Development*, vol. 8, no. 6, November-December 2024, pp. 493-502.

[4] P. Singh, "A layered approach to cybersecurity: People, processes, and technology- explored & explained," July 2021,

https://www.linkedin.com/pulse/layered-approach-cybersecurity-people-processes-singh-casp-cisc-ces

[5] M. Loi et al., "Cybersecurity in health – disentangling value tensions," *Journal of Information, Communication and Ethics in Society*, June 2019,

https://www.emerald.com/insight/content/doi/10.1108/JICES-12-2018-0095/full/html

[6] M. Adams, "Unlocking the benefits of ethical hacking: The importance of ethical hackers in cybersecurity," April 2023,

https://www.businesstechweekly.com/cybersecurity/network-security/ethical-hacking/

[7] M. N. O. Sadiku, S. Alam, S. M. Musa, and C. M. Akujuobi, "A primer on cybersecurity," *International Journal of Advances in Scientific Research and Engineering*, vol. 3, no. 8, Sept. 2017, pp. 71-74.

[8] M. N. O. Sadiku, M. Tembely, and S. M. Musa, "Smart grid cybersecurity," *Journal of Multidisciplinary Engineering Science and Technology*, vol. 3, no. 9, September 2016, pp.5574-5576.

[9] “FCC small biz cyber planning guide,”

https://transition.fcc.gov/cyber/cyberplanner.pdf

[10] “The 8 most common cybersecurity attacks to be aware of,”

https://edafio.com/blog/the-8-most-common-cybersecurity-attacks-to-be-aware-of/

[11] Y. Li and Q. Liu, “A comprehensive review study of cyber-attacks and cyber security; Emerging trends and recent developments,” *Energy Reports*, vol. 7, November 2021,

https://www.sciencedirect.com/science/article/pii/S2352484721007289

[12] Y. Zhang, “Cybersecurity and reliability of electric power grids in an interdependent cyber-physical environment,” Doctoral Dissertation, University of Toledo, 2015.

[13] “Why is cyber security important for telco companies?”

https://bilginc.com/en/blog/why-is-cyber-security-important-for-telco-companies-5550/

[14] D. K. Ali, “Use of cybersecurity in telecom industries,” August 2022,

https://insidetelecom.com/cybersecurity-in-telecom-industries/

[15] “Cybersecurity in the telecom industry: Challenges and career opportunities,”

Jun 2024,

https://www.coursera.org/articles/cyber-security-in-telecom-industry

[16] “Telecom industry security frameworks: Protecting communication networks,”

https://www.neumetric.com/telecom-industry-security-frameworks/#:~:text=Security%20frameworks%20serve%20as%20blueprints,standards%20to%20mitigate%20risks%20%26%20vulnerabilities.

[17] M. N. O. Sadiku, *Cybersecurity and Its Applications*. Moldova, Europe: Lambert Academic Publishing, 2023.

[18] P. Traynor, P. McDaniel, and T. La Porta, *Security for Telecommunications Networks*. Springer, 2008.

[19] N. Boudriga, *Security of Mobile Communications. Boca Raton*, FL: CRC Press, 2009.

[20] A. Jamal, H. Jahankhani, and S. Lawson (eds.), *Cybersecurity, Privacy and Freedom Protection in the Connected World: Proceedings of the 13th International Conference on Global Security, Safety and Sustainability*, London, January 2021. Springer, 2021.

[21] US Congress, *The Future of Cyber and Telecommunications Security at DHS*. U.S. Government Printing Office.

INDEX

www.ingramcontent.com/pod-product-compliance
Lightning Source LLC
LaVergne TN
LVHW050910210926
848482LV00012B/394
9798991226660